AF342085

History of Research on Tumor Angiogenesis

Domenico Ribatti

History of Research
on Tumor Angiogenesis

 Springer

Prof. Domenico Ribatti
Università di Bari
Policlinico
Dipartimento di Anatomia Umana e
Istologia
Bari
Italy
ribatti@anatomia.uniba.it

ISBN 978-1-4020-9559-7   e-ISBN 978-1-4020-9563-4

DOI 10.1007/978-1-4020-9563-4

Library of Congress Control Number: 2008940585

Printed on acid-free paper

9 8 7 6 5 4 3 2 1

springer.com

Dedicated to the memory of Judah Folkman, a pioneer in the study of tumor angiogenesis

Contents

Chapter 1

1.1 Early Evidence of the Vascular Phase and Its Importance in Tumor Growth

The term angiogenesis, meaning the formation of new blood vessels from pre-existing ones, had been coined in 1794 by the British surgeon John Hunter to describe blood vessel growth in reindeer antlers as a result of long-lasting exposure to cold (Hunter, 1794). Studies over the past 30 years have provided significant insights into the angiogenic process and its role in cancer biology with over 17,000 papers published on the topic.

Virchow, the founder of pathological anatomy, drew attention to the huge number of blood vessels in a tumor mass and recognized that the stroma of tumors has a distinctive capillary network as long ago as 1865 (Fig. 1.1). Tumor vascularization was first studied systematically by Goldman (1907), who, by using intra-arterial injections of bismuth in oil, described the vasoproliferative response of the organ in which a tumor develops as follows: "The normal blood vessels of the organs in which the tumor is developing are disturbed by chaotic growth, there is a dilatation and spiralling of the affected vessels, marked capillary budding and new vessel formation, particularly at the advancing border."

The vascular anatomy of tumors was studied in detail by anatomists throughout the nineteenth century. In 1927, Lewis described in detail the vasculature of several tumors that spontaneously occur in rats and observed that the vascular architecture of each tumor type was different, leading to the conclusion that tumor environment has a significant influence on the growth and morphological characteristics of the blood vessels. The structurally diverse nature of neoplastic growth leads to a wide variation in the organization of tumor vasculature and a common pattern cannot be recognized.

In 1928, Sandison introduced the use of a transparent chamber that could be inserted into the rabbit's ear allowing microscopic observation of living tissues underneath a glass coverslip.

When Clark et al. (1931, 1932, 1939) perfected the implantation of transparent chambers in a rabbit's ear, the morphological characteristics of blood and lymphatic vessels could be studied in vivo, including the use of contrast media.

D. Ribatti, *History of Research on Tumor Angiogenesis*,
DOI 10.1007/978-1-4020-9563-4_1, © Springer Science+Business Media B.V. 2009

Fig. 1.1 A portrait of Rudolf Virchow

1.2 Early Evidence of Tumor Cells Releasing Specific Growth Factor for Blood Vessels

In 1939, Ide et al. used transparent chambers to investigate the correlation between growth of a transplanted rabbit carcinoma and vascular supply. They observed that tumor growth is accompanied by the rapid and extensive formation of new vessels and pointed out that if blood vessel growth did not occur, the transplanted tumor failed to grow and they were the first to suggest that tumors release specific factors capable of stimulating the growth of blood vessels.

In 1941, Green demonstrated that the growth of H-31 rabbit carcinomas transplanted into the anterior chamber of the guinea pig did not vascularize and failed to grow for 16–26 months. The transplants remained at a size of approximately 2.5 mm in diameter for the duration of the experiment. However, when the same tumors were reimplanted into the eyes of their original host, they vascularized and expanded in mass, filling the anterior chamber within 50 days.

In 1945, Algire and Chalkley were the first to appreciate that growing malignancies could continuously elicit new capillary growth from the host. They, for the first time, introduced a quantitative approach to assess blood vessel growth by performing daily counts of blood vessels and comparing them to the tumor size. They used a transparent chamber implanted in a cat's skin to study the vasoproliferative reaction secondary to a wound or implantation of normal or neoplastic tissues and showed that the vasoproliferative response induced by tumor tissues was more substantial and earlier than that induced by normal tissues or following a wound. They concluded that the growth of a tumor is closely connected with the development of an intrinsic vascular network and stated that "It is entirely possible that the change in the tumour cell that enables it to evoke capillary proliferation is the only change necessary to give the tumor cell its increased autonomy of growth relative to the normal cell from which it arose."

In his treatise "Il Cancro," published by Ambrosiana in 1946 (Fig. 1.2), Pietro Rondoni, professor of general pathology in the University of Milan and director of the Milan Cancer Institute, stated with regard to the stroma of tumors that "A tumor acts both angioplastically and angiotactically, in other words it promotes the formation of new vessels and attracts vascular outgrowths (capillaries and pluripotent perivasal cells) so as to build up and shape a stroma of its own, a newly formed stroma. It must thus be unreservedly admitted that tumors are partly vascularized by the already existing network of vessels around them. As in other pathological processes, therefore, such neoformation as takes place is a vascular neoformation from budding of the existing capillaries."

The importance of this passage lies in the fact that Rondoni refers to the ability of a tumor to induce the formation of new blood vessels from those that already surround it. He also asserts that this angiogenic activity occurs in its stroma. The topicality of Rondoni's remarks is evident. He was speaking of both the angiogenic capacity of a tumor and the importance of the stroma in new vessel formation. The concept of the context as a microenvironment within which angiogenesis occurs is particularly topical. The context, indeed, appears to govern the time and space patterns of angiogenesis. It also determines whether it will remain confined within physiological bounds or progress to a pathological state and is thus a therapeutical target through which normality may be restored.

In 1948, Michaelson proposed that a diffusible "factor X" produced by the retina was responsible for retinal and iris neovascularization that occurred in proliferative diabetic retinopathy.

In 1956, Merwin and Algire found that the vasoproliferative response of normal or neoplastic tissues transplanted into muscle was not significantly different with respect to the time of onset of new blood vessels, though it was stronger when the implantation was performed in a resection area. In addition, while normal tissues induced a vasoproliferative response confined to the host, tumor tissues induced the formation of neovessels that pierced the implant. Lastly, the intensity of the response seemed to be influenced by the distance between the implant and the host's vessels: normal tissue was unable to induce a response if placed more than 50 μm away, whereas tumor tissue had a longer activity range.

Greenblatt and Shubik (1968) implanted Millipore chambers (pore size 0.45 μm) into a hamster's cheek pouch and placed some tumor fragments around them. In a few days, the growing tumor mass engulfed the whole chamber, whose pores were permeable into the tumor interstitial fluid, but not into the tumor cells. New blood vessels, however, were formed in any case very likely through the release

Fig. 1.2 The cover of the treatise "Il Cancro", written by the Italian scientist Pietro Rondoni

of a diffusible factor that could pass through the pores. Ehrman and Knoth (1968) confirmed these data with tumor fragments laid on Millipore filters planted on the chick embryo chorioallantoic membrane (CAM).

In 1968, Tannock further explored the relationship between endothelial cell and tumor cell proliferation in a transplantable mammary gland carcinoma, using autoradiographic techniques. He showed that the mitotic index of tumor cells decreases with increased distance from endothelial cells, providing direct evidence that diffusion of oxygen and nutrients from the vascular endothelium is a rate-limiting step in tumor cell growth.

The cause of tumor neovascularization in these studies was attributed variously to inflammation, vasodilation, increased tumor metabolism, overproduction of specific metabolites such as lactic acid, or to hypoxia.

1.3 Tumors in Isolated Perfused Organs: Absence of Angiogenesis

In 1963, Folkman and collaborators were studying hemoglobin solutions as potential substitutes for blood transfusion. To test which solution was optimal for tissue survival, they perfused these solutions through the vasculature of canine thyroid glands, by using an apparatus with a silicone rubber oxygenator (Fig. 1.3). The glands survived for about 2 weeks. They could distinguish different hemoglobin preparations by histologic analysis of the thyroid glands after a week or more of continuous arterial perfusion. To determine if these isolated organs could support growth, they injected them with adult mouse melanoma cells. Tiny tumors developed but stopped growing at 1–2 mm diameter and never became vascularized (Folkman et al., 1963). When isolated thyroid glands were perfused with platelet-rich plasma, endothelial vascular integrity was preserved in the isolated perfused organs, whereas in organs perfused with platelet-poor plasma, endothelial cells were disrupted within 5 h (Gimbrone et al., 1969). However, the tumors were not dead. When they were transplanted to syngeneic mice, they rapidly vascularized and grew to more than $1\,cm^3$, more than 1,000 times their original volume in the perfused thyroid gland (Fig. 1.3).

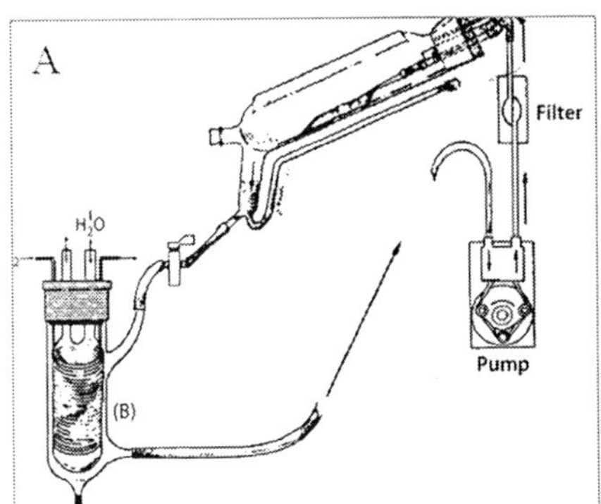

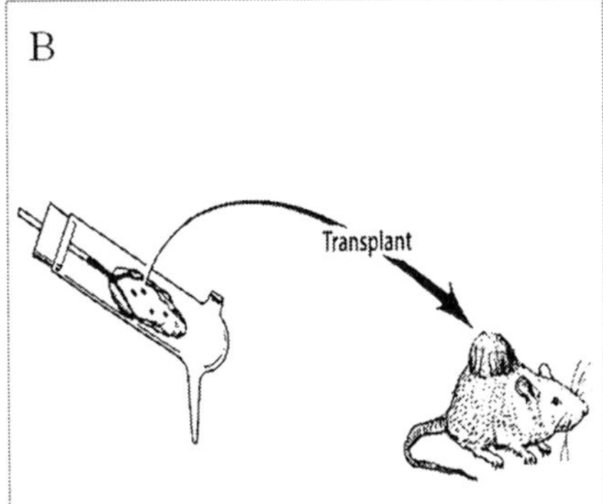

Fig. 1.3 Perfusion of isolated canine thyroid gland through the carotid artery with hemoglobin solution. (**A**) The perfusion circuit includes a silicon rubber oxygenator and a roller pump with silicon rubber tubing. (**B**) When the non-expanding tumor was transplanted to a syngeneic mouse, it grew more than 1,000 times its initial volume in the perfused thyroid gland. (Reproduced with modifications from Folkman J, Tumor angiogenesis: from bench to bedside, in "Tumor Angiogenesis", Marmé D. and Fysening N, eds., Springer, Berlin 2008, pp. 3–28.)

This was the first evidence to show that the absence of neovascularization corre-lated with severe restriction of tumor growth.

1.3.1 Hypothesis: Tumor Growth Is Angiogenesis-Dependent

In 1971, Folkman published in the *New England Journal of Medicine* a hypothesis that tumor growth is angiogenesis-dependent and that inhibition of angiogenesis could be therapeutic (Folkman, 1971). This paper also introduced the term anti-angiogenesis to mean the prevention of new vessel sprout from being recruited by a tumor. The hypothesis predicted that tumors would be enabled to grow beyond a microscopic size of 1–2 mm^3 without continuous recruitment of new capillary blood vessels (Fig. 1.4).

This concept is now widely accepted because of supporting data from experi-mental studies and clinical observations carried out over the intervening years.

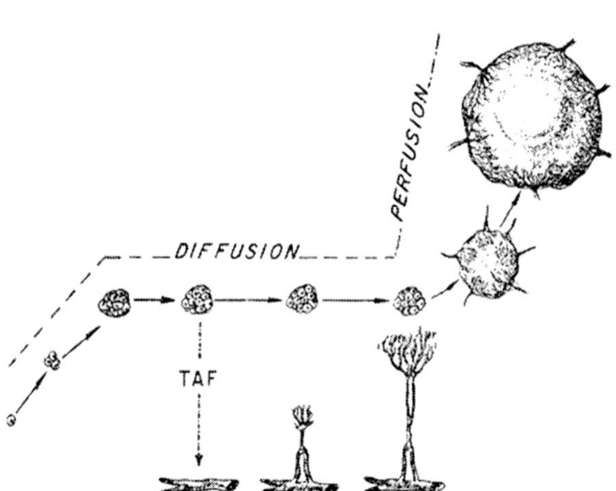

Fig. 1.4 Schematic drawing showing that most solid tumors may exist early as tiny cell populations living by simple diffusion in the extracellular space. Further growth requires neovascularization and tumor angiogenesis factor (TAF) may be the mediator of neovascularization. (Reproduced from Folkman J, *N Engl J Med*, 285: 1182–6, 1971.)

1.3.2 Evidence that Tumors Are Angiogenesis-Dependent

Folkman and collaborators provide evidence for the dependence of tumor growth on neovascularization:

(1) Tumor growth in the avascular cornea proceeds slowly at a linear rate, but after vascularization, tumor growth is exponential (Gimbrone et al., 1974).

(2) Tumors suspended in the aqueous fluid of the anterior chamber of the rabbit's eye and observed for a period of up to 6 weeks remain viable, avascular, and of limited size (less than 1 mm^3) and contain a population of viable and mi-totically active tumor cells. These tumors induce neovascularization of the iris vessels, but are too remote from these vessels to be invaded by them. After

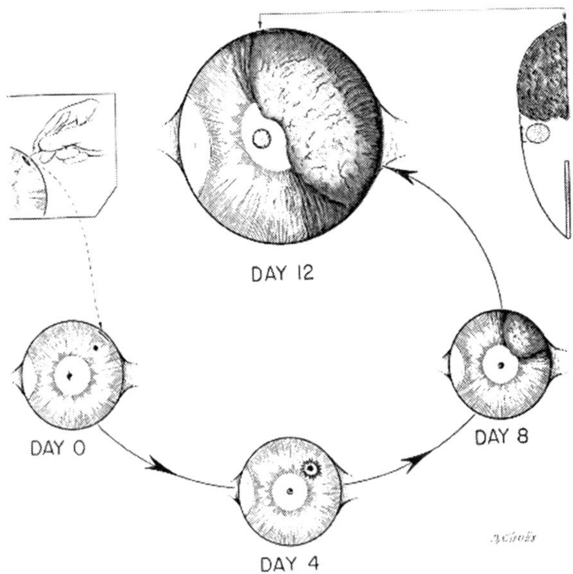

Fig. 1.5 The patterns of development of two simultaneous implants of Brown-Pearce tumor in the rabbit's eye. The anterior chamber implant remains avascular, while the iris implant vascularizes and grows progressively. (Reproduced from Gimbrone MA Jr. et al., *J Exp Med*, 136: 261–76, 1972.)

implantation contiguous to the iris, which had abundant blood vessels, the tumors induced neovascularization and grew rapidly, reaching 16,000 times the original size within 2 weeks (Fig. 1.5) (Gimbrone et al., 1972). This experiment introduced the concept of tumor dormancy brought about by the prevention of neovascularization. During the dormancy period, the tumors developed a central necrotic core surrounded by a layer of viable cells, in which mitotic figures were observable. Overall, these microscopic tumors remained avascular, as demonstrated by using microscopic and histological analyses and fluorescein tests. In a parallel study tumors were suspended in the aqueous humor of the anterior chamber, placed at various distances from the iris vessels, and compared with tumors implanted directly on the iris and with those implanted in the cornea (Gimbrone et al., 1973). Moving the distant, dormant tumors closer to the iris jump-started their growth. This suggested that this type of tumor dormancy was caused not by cell cycle arrest or immune control, but by a lack of blood supply. These experimental data were in accord with clinical evidence showing a large retinoblastoma in the eye (>1 cm^3), protruding from the retina into the vitreous (Folkman, 1975). It was highly vascularized. Its metastases in the vitreous and the aqueous humor were not neovascularized because they were floating at too great a distance from the nearest vascular bed. When cryotherapy was used to regress retinoblastoma, the tiny metastases fell on the vascular bed vacated by the primary tumor, and became neovascularized themselves.

(3) B-16 mouse melanoma, V-79 Chinese hamster lung cells, and L-5178 Y murine leukemia cells were plated in soft agar (Folkman and Hochberg, 1983). After 6–7 days of incubation, spheroid colonies of 0.1 mm were visible. All spheroids first enlarged exponentially for a few days and then continued on a linear growth

curve for 5–23 weeks before reaching a diameter beyond which there was no further expansion. This was termed the dormant phase. After the dormant diameter was reached, these spheroids remained viable for 3–5 months, or as long as they were frequently transferred to a new medium. Cells in the periphery of the spheroid incorporated ^{3}H-labeled thymidine while cells in the center died. This is a form of population dormancy in which the proliferating cells near the surface of the spheroid just balance those dying cells deep in the center of the spheroid.

(4) Tumors grown in the vitreous of the rabbit's eye remain viable but attain diameters of less than 0.50 mm for as long as 100 days. Once such a tumor reaches the retinal surface, it becomes neovascularized and within 2 weeks can undergo a 19,000-fold increase in volume over the avascular tumor (Brem et al., 1976).

(5) The CAM appears at day 5 during the development of the chick embryo. The ^{3}H-thymidine labeling index of its vascular endothelium decreases with age, with an abrupt reduction at day 11 (Ausprunk et al., 1974). Prior on 11 day, labeling index is approximately 23%; at 11 days, the labeling index decreases to 2.8%, and subsequently the cells begin to acquire the structural characteristics of matured, differentiated endothelium. One-millimeter fragments of fresh Walker 256 carcinoma were implanted on the CAM from day 3 to day 16 (Knighton et al., 1977). The size of the tumors was measured daily, and the onset of vascularization of each tumor was determined in vivo with a stereomicroscope and confirmed with histological sections. Proliferation of chick capillaries occurred in the neighborhood of the tumor graft within 24 h after implantation, but capillary sprouts did not penetrate the tumor graft until approximately 72 h. During the avascular phase, tumor diameter did not exceed 1 mm. Small tumor implants of 0.5 mm or less grew to 1 mm and stopped expanding. Larger tumor implants of 2 or 3 mm shrank until they reached 1 mm diameter. During the first 24 h after penetration by capillaries, there was a rapid tumor growth. Neovascularization was not grossly observable with the stereomicroscope until after day 10 or 11. Tumors implanted on the CAM after day 11 grew at a slower rate in parallel with the reduced rates of endothelial growth.

(6) When tumor grafts of increasing size (from 1 to 4 mm) are implanted on the 9-day CAM, grafts larger than 1 mm undergo necrosis and autolysis during the 72-h prevascular phase. They shrink rapidly until the onset of neovascularization, when rapid tumor growth resumes (Knighton et al., 1977). In another study (Ausprunk et al., 1975) the behavior of tumor grafts on the CAM was compared to grafts of normal adult and embryonic tissues. In tumor tissue, pre-existing blood vessels within the tumor graft disintegrated by 24 h after implantation. Neovascularization did not occur until after at least 3 days, and only by penetration of proliferating host vessels into the tumor tissue. There was marked neovascularization of host vessels in the neighborhood of the tumor graft. By contrast, in embryonic graft, pre-existing vessels did not disintegrate. They are reattached by anastomosis to the host vessels within 1–2 days, but with minimal or almost no neovascularization on the part of the host vessels. In adult tissues, the pre-existing graft vessels disintegrated, there was no reattachment of their

circulation with the host, and adult tissues did not stimulate capillary proliferation. These studies suggest that only tumor grafts are capable of stimulating formation of new blood vessels in the host.

(7) In transgenic mice that develop carcinomas of the β cells in the pancreatic islets, large tumors arise only from a subset of preneoplastic hyperplastic islets that have become vascularized (Folkman et al., 1989).

1.4 The Avascular and Vascular Phases of Solid Tumor Growth

Solid tumor growth consists of an avascular and a subsequent vascular phase. Assuming that it is dependent on angiogenesis and that this depends on the release of angiogenic factors, acquisition of angiogenic capability can be seen as an expression of progression from neoplastic transformation to tumor growth and metastasis.

In the 1970s, using the rodent mammary gland as a model, Gullino and coworkers observed that adult resting mammary gland has limited, if any, angiogenic capacity. However, this is consistently acquired by mammary carcinomas. Interestingly, lesions with high frequency of neoplastic transformation induced angiogenesis at a much higher rate than did lesions with low frequency of transformation. This elevated angiogenic capacity was observed long before any morphological sign of neoplastic transformation (Gimbrone and Gullino, 1976a,b; Brem et al., 1977; 1978; Maiorana and Gullino, 1978). Hyperplastic lesions of the human mammary gland showed a similar behavior (Gimbrone and Gullino, 1976a). Thus, angiogenesis may represent an early marker for neoplastic transformation. After several years, the established role of oncogene activation and oncosuppressor gene inactivation in modulating the expression of pro- and anti-angiogenic factors has confirmed Gullino's observations at a molecular level (Rak et al., 2002).

The avascular phase appears to correspond to the histopathological picture presented by a small colony of neoplastic cells (500,000 to 1 million cells/1–2 mm in diameter) that reaches a steady state before it proliferates and becomes rapidly invasive. Here, metabolites and catabolites are transferred by simple diffusion through the surrounding tissue. The cells at the periphery of the tumor continue to reproduce, whereas those in the deeper portion die away.

1.4.1 First Evidence of the Existence of the Avascular and Vascular Phases of Solid Tumor Growth

The earliest evidence of the existence of the two phases was obtained by Folkman and collaborators in 1963, who perfused the lobe of a thyroid gland with plasma and inoculated a suspension of melanoma B-16 tumor cells through the perfusion fluid. These cells grew into small, clearly visible black nodules. The nodules did not exceed 1 mm in diameter and did not connect with the host's vascular network. Their outer third generally remained vital, while the interior portion underwent necrosis.

Reimplanted nodules, on the other hand, equipped themselves with a vascular network and grew very rapidly. The conclusion was thus drawn that the absence of vascularization limits the growth of solid tumors.

Further research by Folkman's group resulted in an experimental system in which the tumor, or its extracts, could be separated from the vascular bed (Cavallo et al., 1972; 1973). This system was based on subcutaneous insufflation to lift the skin of a rat and form a poorly vascularized region below it. Millipore filters containing Walker 256 cancer cells or their cytoplasmic or nuclear extracts (TAF) were implanted into the fascial floor of the dorsal air sac. At intervals thereafter, ^{3}H-labeled thymidine was injected into the air sac and the tissues were examined by autoradiography and electron microscopy. Autoradiographs showed thymidine-^{3}H labeling in endothelial cells of small vessels, 1–3 mm from the site of implantation, as early as 6–8 h after exposure to tumor cells. DNA synthesis by endothelium subsequently increased, and within 48 h new blood vessels formation was detected. The presence of labeled endothelial nuclei, endothelial mitosis, and regenerating endothelium was confirmed by electron microscopy. TAF also induced neovascularization and endothelial DNA synthesis after 48 h. Further ultrastructural autoradiographic studies were carried out with the same model (Cavallo et al., 1973). It was apparent that by 48 h there was ultrastructural evidence of regenerating endothelium, including marked increase in ribosomes and endoplasmic reticulum, scarce or absence of pinocytotic vesicles, and discontinuous basement membrane. Labeled endothelial cells were seen along newly formed sprouts as well as in parent vessels. Furthermore, pericytes were also shown to synthesize DNA.

In another series of experiments, 1-mm fragments from Brown-Pearce and V2 carcinomas were implanted into the avascular stroma of a rabbit cornea 1–6 mm away from the limbic vessels, and the tumor growth was observed daily with a stereomicroscope (Gimbrone et al., 1974). After 1 week, new blood vessels had invaded the cornea starting from the edge closer to the site of implantation and developed in that direction at 0.2 mm and then about 1 mm/day. Once the vessels reached the tumor, it grew very rapidly to permeate the entire globe within 4 weeks.

1.4.2 The Significance of Angiogenesis in Hematological Malignancies

There has been an increased interest in recent years in the role of angiogenic cytokines and their receptors in hematological malignancies (Table 1.1). The purpose of this interest is to develop anti-angiogenic drugs that are potentially less toxic than traditional chemotherapeutic drugs.

In 1994, Vacca et al. demonstrated for the first time that bone marrow microvascular density was significantly increased in multiple myeloma (MM) compared to monoclonal gammopathy of undetermined significance (MGUS) and in active versus non-active myeloma and first hypothesized that progression from MGUS to

Table 1.1 Historical reviewing of angiogenesis involvement in hematological malignancies

Angiogenesis	Reference
First evidence of bone marrow angiogenesis in MM High correlation between the extent of bone marrow angiogenesis and plasma cell proliferation The role of the "vascular phase" in disease progression	Vacca et al. (1994)
Role of bone marrow microenvironment in MM	Klein et al. (1995)
First evidence of bone marrow angiogenesis in B-NHL High correlation between the extent of bone marrow angiogenesis and B-NHL grading	Ribatti et al. (1996)
First evidence of increased bone marrow microvessel density in ALL	Perez-Atayde et al. (1997)
Role of mast cells tryptase-positive in angiogenic cascade in B-NHL, MM and B-CLL	Ribatti et al. (1998, 1999a, 2003a)
Induction of angiogenesis by plasma cells secretion of FGF-2 and MMP-2 in active MM	Vacca et al. (1999)
First evidence of angiogenesis involvement in the pathogenesis of B-CLL	Molica et al. (1999)
High bone marrow and serum levels of angiogenic cytokines in MM	Di Raimondo et al. (2000)
High expression of VEGF in plasma cells, myeloid and monocyte precursors	Bellamy et al. (2001)
High synthesis of MMP-2 and MMP-9 by B-CLL cells	Bauvois et al. (2002)
Detailed phenotypic, genetic, and functional characterization of bone marrow endothelial cells from patients with MM	Vacca et al. (2003)
Demonstration of a pre-angiogenic phase in MM characterized by high amount of CD45-cells	Asosingh et al. (2004)

MM: multiple myeloma; B-NHL: B cell non-Hodgkin's lymphomas; ALL: acute lymphocytic leukemia; B-CLL: B cell chronic lymphocytic leukemia.

myeloma is accompanied by an increase in bone marrow microvascular density and that microvascular density is related to the plasma cell labeling index.

Increased angiogenesis has also been demonstrated in acute and chronic lymphocytic leukemia (ALL and CLL), acute myeloid leukemia (AML), as well as lymphomas and has been found to have prognostic value.

1.4.2.1 Angiogenesis in Multiple Myeloma

Assuming that in MM, microvascular density depends on angiogenesis, these results are consistent with the notion that angiogenesis favors expansion of the MM mass by promoting plasma cell proliferation. Myeloma plasma cells induce angiogenesis directly via the secretion of angiogenic cytokines, such as vascular endothelial growth factor (VEGF) and fibroblast growth factor-2 (FGF-2), and indirectly by the induction of host inflammatory cell infiltration, and degrade the extracellular matrix with their matrix-degrading enzymes, such as matrix metalloproteinases-2

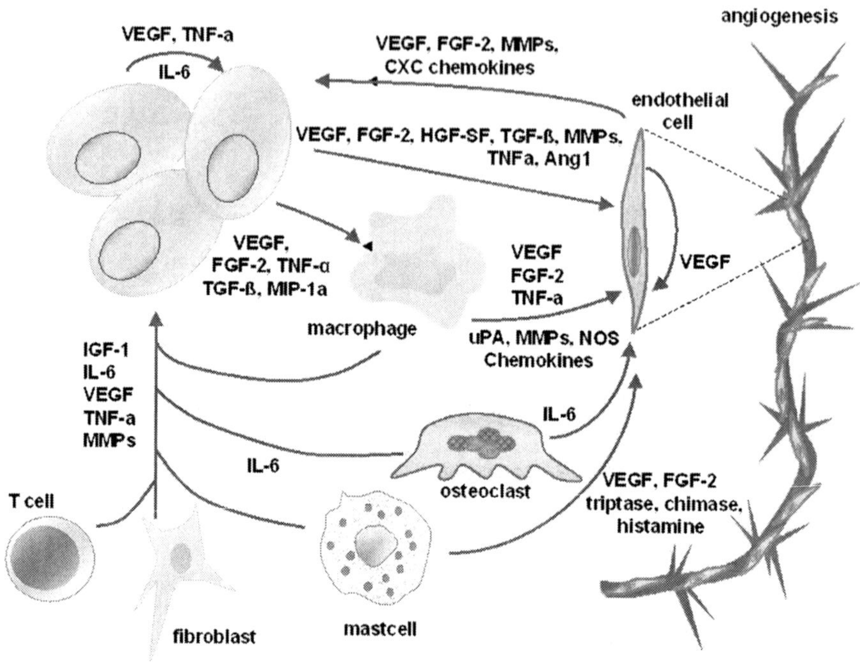

Fig. 1.6 Interplay between various microenvironmental cells and factors promoting angiogenesis in multiple myeloma. (Reproduced from Ribatti D et al., *Oncogene*, 25: 4257–66, 2006.)

and -9 (MMP-2 and MMP-9) and urokinase-type plasminogen activator (Vacca and Ribatti, 2006).

Reciprocal positive and negative interactions between plasma cells and bone marrow stromal cells, namely, hematopoietic stem cells, fibroblasts, osteoblasts/osteoclasts, chondroclasts, endothelial cells, endothelial cell progenitor cells, T cells, macrophages, and mast cells, mediated by an array of cytokines, receptors, and adhesion molecules, modulate the angiogenic response in MM (Ribatti et al., 2006a) (Fig. 1.6). Macrophages and mast cells contribute to build neovessels in active MM through vasculogenic mimicry, and this ability proceeds parallel to the progression of the plasma cell tumors (Figs. 1.7, 1.8) (Scavelli et al., 2008; Nico et al., 2008a).

1.4.2.2 Angiogenesis in Leukemia

In a study of 51 children with ALL, microvascular density in bone marrow increased 6–7-fold compared with the control bone marrows of children evaluated for primary tumor (Perez-Atayde et al., 1997). Urinary levels of FGF-2 were high in these patients before induction therapy, variable during induction, and normalized when a complete response was achieved (Perez-Atayde et al., 1997). Aguayo et al. (2000) have provided further evidence of increased bone marrow microvascular density in ALL as well as increased plasma levels of FGF-2, but not VEGF. These investigators

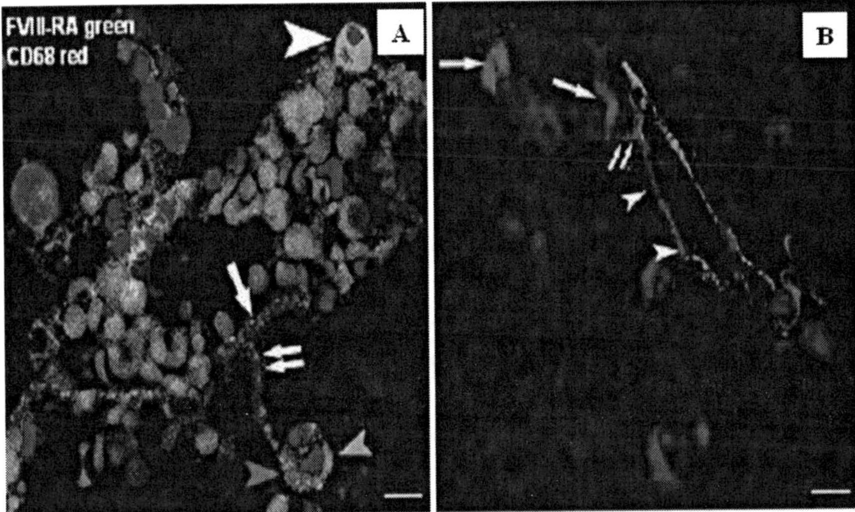

Fig. 1.7 (**A**) Dual confocal laser microscopy of a microvessel lined by flattened FVIII$^+$ multiple myeloma endothelial cells (*arrow*), an FVIII$^+$ macrophage (*arrowhead*) showing protrusions connected to multiple myeloma endothelial cells, and another macrophage containing double-labeled CD68 (*arrowhead*) and FVIII (*arrowhead*) granules in the cytoplasm and connected to multiple myeloma endothelial cells by an FVIII$^+$ cytoplasmic protrusion (*double arrow*). Erythrocytes are well-recognizable in the lumen. (**B**) Another microvessel formed by FVIII$^+$ multiple myeloma endothelial cells and CD68$^+$ (*arrowheads*) tracts that belong to the cytoplasmic protrusions (*double arrows*) of macrophages, some of which are arrowed. (Reproduced from Scavelli C et al., *Oncogene*, 27: 663–74, 2008.)

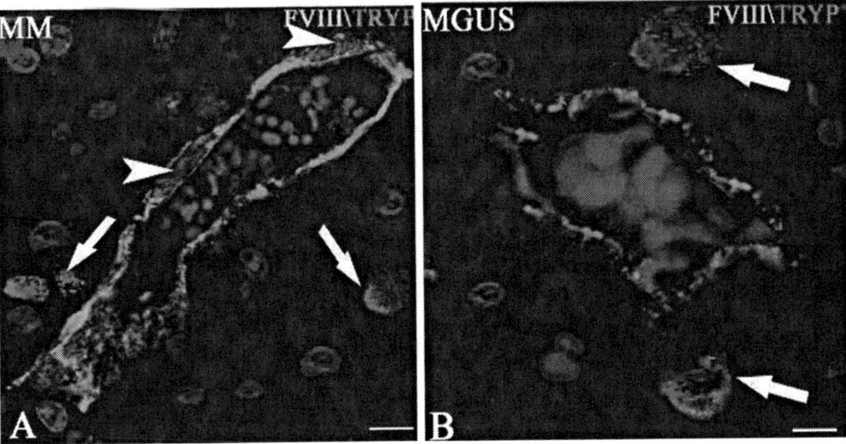

Fig. 1.8 Double FVIII and tryptase confocal laser microscopy from multiple myeloma (MM) (**A**) and monoclonal gammopathies of undetermined significance (MGUS) (**B**) of bone marrow biopsy specimens. (**A**) An MM vessel is lined by both endothelial cells FVIII$^+$ and by mast cells tryptase$^+$ (*arrowheads*). Mast cells containing tryptase$^+$ granules (*arrows*) are also recognizable on the abluminal side of the vessel. (**B**) An MGUS vessel is lined only by endothelial cells FVIII$^+$ and is surrounded by tryptase$^+$ mast cells (*arrows*). (Reproduced from Nico B et al., *Stem Cells Dev*, 17: 19–22, 2008.)

also demonstrated that intracellular levels of VEGF in leukemic blasts of patients with acute myelogenous leukemia were higher than in mononuclear cells from controls (Aguayo et al., 1999). Increasing VEGF levels were associated with shorter overall and disease-free survival, but did not correlate with established prognostic factors, such as blast counts, age, and cytogenetic abnormalities.

Angiogenesis may also be involved in the pathogenesis of B-cell CLL (Letilovic et al., 2006). High microvascular density has been found in CLL and is associated with poor prognosis (Molica et al., 2002). Though the role of angiogenesis in the pathophysiology of CLL remains to be fully elucidated, experimental data suggest that several angiogenic factors, such as VEGF and FGF-2, play a role in disease progression (Letilovic et al., 2006). The coexpression of angiogenic molecules and their receptors suggests that the biology of the leukemic cells might also be directly impacted by angiogenic factors as a result of autocrine pathways of stimulation. Additionally, interactions between CLL B-cells and their microenvironment generate alterations in the secretion of angiogenic factors that result in enhanced leukemic B-cell resistance to apoptotic cell death (Shanafelt and Kay, 2006).

1.4.2.3 Angiogenesis in Acute Myeloid Leukemia (AML)

AML is associated with an increased bone marrow angiogenesis. Microvascular density significantly decreases after induction chemotherapy and angiogenesis correlates with clinical outcome of patients, whereas there is a persistence of increased microvascular density in patients with residual leukemic blast infiltration after induction chemotherapy (Rabtish et al., 2004). Increased expression levels of VEGF and FGF-2 are found in the plasma from patients with AML compared with that in the control group. Moreover, AML cells not only synthesize VEGF but also express functional VEGF receptors (VEGFRs), resulting in autocrine loop for tumor growth and progression. High pretherapeutic levels of angiopoietin-2 (Ang-2) in the bone marrow indicate a favorable prognosis in AML patients treated with chemotherapy (Loges et al., 2005).

1.4.2.4 Angiogenesis in Lymphomas

Angiogenesis and angiogenic factors are increased in most lymphomas and are associated with an adverse outcome or more aggressive behavior in malignant lymphoma (Koster and Raemackers, 2005). Ribatti et al. (1996) showed increased microvascular density in non-Hodgkin's lymphoma (NHL). High-grade lymphomas had higher counts than intermediate-grade lymphomas, which in turn had higher counts than low-grade lymphomas, suggesting that angiogenesis may be involved in disease progression. Vacca et al. (1997) documented higher microvascular density in mycosis fungoides lesions compared to normal skin, with an increase in microvascular density with disease progression.

High levels of VEGF in blood and tissue are associated with an adverse prognosis and structural microvessel abnormalities are present in some lymphoma subtypes. However, given that malignant lymphoma is a clinically and histologically heterogeneous group of diseases, the role of angiogenesis is likely to differ between the various lymphoma subtypes (Koster and Raemackers, 2005).

1.4.3 Non-Angiogenesis-Dependent Pathways for Tumor Growth: Vascular Cooption and Vasculogenic Mimicry

Pezzella et al. (1997) were the first to describe a non-small-lung carcinoma growing with no morphological evidence of neoangiogenesis by exploiting normal tissue vessels. They reported that lung carcinomas without angiogenesis are characterized by lack of parenchymal destruction and absence of new vessels and tumor-associated stroma. They also questioned whether the neoplastic cells were truly non-angiogenic, as suggested by the apparent lack of new vessel formation. Clinical–pathological study has shown that patients with a putative non-angiogenic carcinoma have more aggressive disease (Pastorino et al., 1997). A pattern of non-angiogenic growth has also been described in glioblastoma multiforme (Wesseling et al., 1995).

It has also been suggested, looking at the microvessel density in the primary tumor and in synchronous nodal metastases, that lymph node is another site in which tumors grow independently of their angiogenic ability (Pezzella, 2000; Guidi et al., 2000; Naresh et al., 2001). Naresh et al. reported that the percentage of endothelial cells in cell cycle is higher in primary tumors than in metastases and suggested that while in the primary tumors a high vascular proliferating fraction is due to angiogenesis, in the nodal metastases a low proliferating fraction in the endothelium indicates a reduced angiogenesis.

Holash et al. (1999a) reported that tumor cells migrate toward existing host organ blood vessels in sites of metastases, or in vascularized organs such as the brain, to initiate blood vessel-dependent tumor growth as opposed to classic angiogenesis. These vessels then regress owing to apoptosis of the constituent endothelial cells, apparently mediated by Ang-2. Ang-2, a ligand for the endothelial tyrosine kinase-receptor Tie-2, antagonizes the activity of the other Tie-2 ligand, Ang-1, that keeps the vessel in a quiescent state by maintaining high pericyte coverage (Holash et al., 1999a). Finally, at the periphery of the growing tumor mass angiogenesis occurs by cooperative interaction of VEGF and Ang-2.

Tumor cells often appear to have immediate access to blood vessels, such as when they metastasize to or are implanted within a vascularized tissue (Holash et al., 1999b; Zagzag et al., 1999). They immediately coopt and often grow as cuffs around adjacent existing vessels. A robust host defense mechanism is activated, in which the coopted vessels initiate an apoptotic cascade, probably by autocrine induction of Ang-2, followed by vessel regression. This is of the coopted vessels that carries off much of the dependent tumor and results in massive tumor death. However, successful tumors overcome this vessel regression by initiating neoangiogenesis.

Shortly after regression, a tumor upregulates its expression of VEGF, presumably because it is becoming hypoxic due to the loss of vascular support. As in normal vascular remodeling, the destabilizing signal provided by Ang-2, which leads to vessel regression in the absence of VEGF, potentiates the angiogenic response in combination with VEGF. Many solid tumors may fail to form a well-differentiated and stable vasculature because their newly formed tumor vessels continue to over-express Ang-2. Ang-2 induction in host vessels in the periphery of experimental C6 glioma precedes VEGF upregulation in tumor cells and causes regression of coopted vessels (Holash et al., 1999a,b; Yancopoulos et al., 2000).

Vajkoczy et al. (2002) have demonstrated parallel induction of Ang-2 and VEGFR-2 in quiescent host endothelial cells, suggesting that their simultaneous activity is critical for the induction of tumor angiogenesis during vascular initiation of micro-tumors. Consequently, the simultaneous expression of VEGFR-2 and Ang-2, rather than the expression of Ang-2 alone, may indicate the angiogenic phenotype of en-dothelial cells and thus provide an early marker of activated host vasculature. The VEGF/Ang-2 balance may determine whether the new tumor vessels continue to expand when the ratio of VEGF to Ang-2 is high, or regress when it is low during remodeling of the tumor microvasculature.

Maniotis et al. (1999) described a new model of formation of vascular channels by human melanoma cells and called it "vasculogenic mimicry" to emphasize the de novo generation of blood vessels without the participation of endothelial cells and independent of angiogenesis. The word "vasculogenic" was selected to indicate the generation of the pathway de novo and "mimicry" was used because tumor cell pathways for transporting fluid in tissues were clearly not blood vessels (Table 1.2).

Maniotis et al. (1999) showed that highly invasive uveal and cutaneous melanoma cells formed looping patterns positive with the periodic acid-Schiff (PAS) stain in three-dimensional cultures on type I collagen and Matrigel, independent of endothe-lial cells and fibroblasts. These PAS-positive patterns were thus thought to con-tribute to a microcirculation in human melanoma and that the pattern-generating ag-gressive melanoma cells may contribute to a limited local extravascular circulation.

Microarray gene chip analysis of a highly aggressive human cutaneous melanoma cell lines compared with a poorly aggressive ones revealed a significant increase in the expression of laminin 5 and MMP-1, -2, and -9 and MT1-MMP in the highly aggressive cells, suggesting that they interact with and alter their extracellular envi-ronment differently than the poorly aggressive cells, and that increased expression

Table 1.2 Examples of vasculogenic mimicry

Melanoma
Breast carcinoma
Prostatic carcinoma
Ovarian carcinoma
Lung carcinoma
Synoviosarcoma
Rhabdomyosarcoma
Pheochromocytoma

of MMP-2 and MT1-MMP along with matrix deposition of laminin 5 is required for their "vasculogenic mimicry."

Electron microscopy has shown that during "vasculogenic mimicry," channels are lined by a layer of extracellular matrix material, which has the appearance of a basement membrane (Maniotis et al., 1999). No cells or nuclei were detected on the luminal side of the basement membrane and an external layer of tumor cells surrounded the extracellular matrix channels.

These data have been vigorously disputed by Mc Donald et al. (2000), who considered the evidence presented to be neither persuasive nor novel. In their opinion, the data are not convincing because three key questions were not addressed: (i) If erythrocytes are used as markers, are they located inside or outside blood vessels? (ii) Where is the interface between the endothelial cells and tumor cells in the blood vessel wall? (iii) How extensive is the presumptive contribution of tumor cells to the lining of blood vessels? Mc Donald and coworkers argued that the structures described as channels by Maniotis and collaborators are laminar septa that separate groups of tumor cells from one another.

Moreover, the possibility that cancer cells participate in the formation of blood vessels in tumors has been recognized for many years. Tumor cells in some uveal melanomas line cavernous spaces or cyst-like blood lakes that communicate with the microvasculature (Jensen, 1976; Duke-Elder and Perkins, 1996; Francois and Neetens, 1967). Warren (1979), Prause and Jensen (1980) and Hammersen et al. (1985) subsequently added ultrastructural evidence of the contribution of cancer cells to the wall of tumor vessels. Tubes lined by tumor cells have been demonstrated histologically in melanoma (Baron et al., 2000), ovarian carcinoma (Sood et al., 2001), and inflammatory breast cancer (Shirakawa et al., 2002).

Another possibility is that the endothelial cell lining is replaced by tumor cells, resulting in the so-called mosaic vessels, where both the endothelial and tumor cells contribute to the formation of vascular tube (Chang et al., 2000). These authors used CD31 and CD105 to identify endothelial cells and endogenous green fluorescent protein (GFP) labeling of tumor cells, and showed that approximately 15% of perfused vessels of a colon carcinoma xenografted at two sites in mice were mosaic with focal regions where no CD31/CD105 immunoreactivity was detected and tumor cells were in contact with the vessel lumen. This formation of mosaic vessels is distinct from vasculogenic mimicry, as described by Maniotis et al. (1999).

Chapter 2

2.1 Isolation of the First Angiogenic Tumor Factor

Until the early 1970s it was widely assumed that tumors did not produce specific angiogenic proteins. The conventional wisdom was that tumor vasculature was an inflammatory reaction to dying or necrotic tumor cells.

Previous studies had shown that tumor-stimulated vessel growth did not require direct contact between tumor and host tissue (Greenblatt and Shubik, 1968; Ehrman and Knoth, 1968). This made sense to Folkman, who reasoned that a soluble factor would be more likely to reach near, than distant blood vessels. He and his colleagues isolated an angiogenic factor in 1971 (Folkman et al., 1971). The homogenate of a Walker 256 carcinoma – a breast tumor of Sprague-Dawley rats – was fractionated by gel-filtration on Sephadex G-100 (Fig. 2.1). The fraction that exhibited the strongest angiogenic activity had a molecular weight of about 10,000 Da and consisted of 25% RNA, 10% proteins, and 58% carbohydrates, plus a possible lipid residue. It was inactivated by digestion with pancreatic ribonuclease or by heating at 56 °C for 1 h, and was neither modified when kept at 4 °C for 3 months, nor when treated with trypsin for more than 3 days. This active fraction was subsequently called "tumor angiogenesis factor" (TAF) (Folkman et al., 1971). Both the cytoplasmic and the nuclear fractions of tumor cells stimulated angiogenesis. In the nuclear fraction, this was found to be associated with non-histonic proteins (Tuan et al., 1973). TAF has since been non-destructively extracted from several tumor cell lines, and several low molecular weight angiogenic factors have been isolated, again from the Walker 256 carcinoma. These factors induced a vasoproliferative response in vivo when tested on rabbit cornea or chick CAM, and in vitro on cultured endothelial cells (Weiss et al., 1979; McAuslan and Hoffman, 1979; Fenselau et al., 1981).

D. Ribatti, *History of Research on Tumor Angiogenesis*,
DOI 10.1007/978-1-4020-9563-4_2, © Springer Science+Business Media B.V. 2009

TABLE I

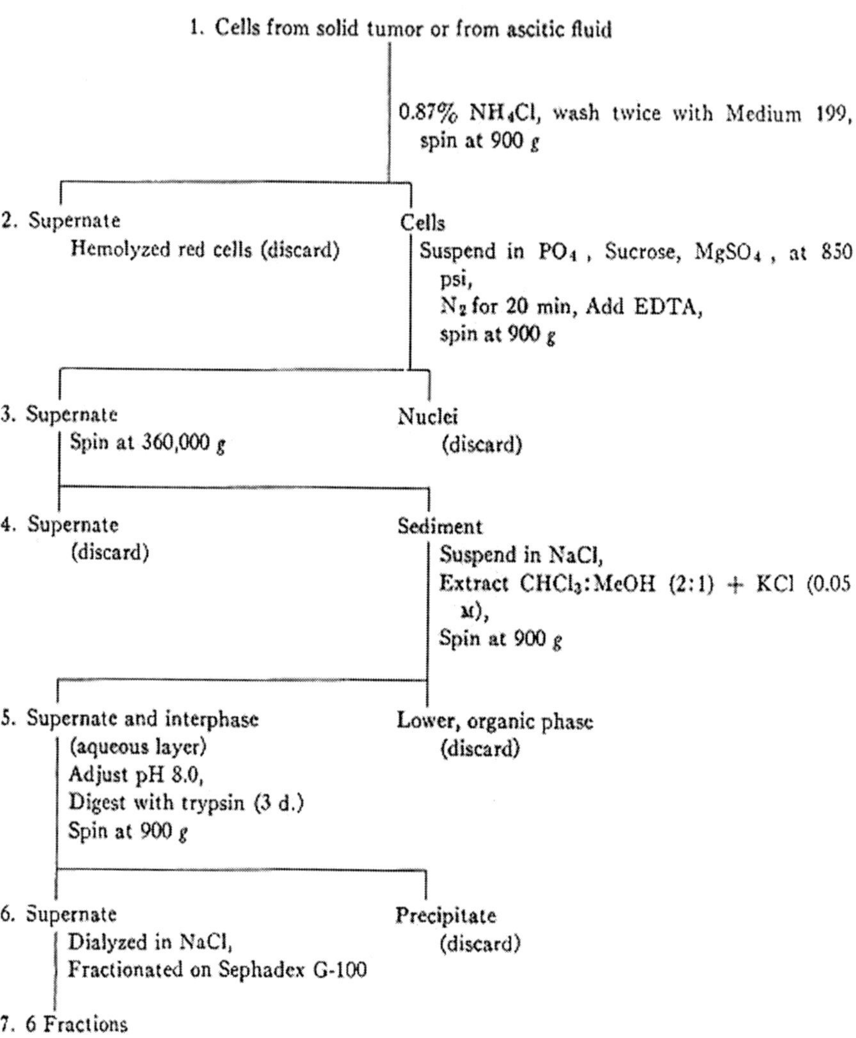

Fig. 2.1 The procedure to isolate tumor angiogenesis factor (TAF). (Reproduced from Folkman J et al., *J Exp Med*, 133: 275–88, 1971.)

2.2 The Discovery of Basic Fibroblast Growth Factor/Fibroblast Growth Factor-2

Gospodarowicz found that the pituitary contains a potent agent for cell growth (Gospodarowicz, 1975). Although he did not identify the factor, he showed that it induces fibroblast cell growth and, hence, he named it fibroblast growth factor (FGF). The mitogenic activity was found to be due to a molecule with a molecular weight of 14,000–16,000 Da and a basic isoelectric point. The activity was not re-stricted to fibroblasts, stimulating many cell types, including endothelial cells and chondrocytes (Gospodarowicz et al., 1978).

In 1984, Shing and co-workers (Shing et al., 1984) reported that they had finally succeeded in isolating and purifying the first factor that specifically stimulated the growth of endothelial cells. This factor bound to heparin with such a high affinity and could be purified 200,000-fold by a single passage over a heparin affinity col-umn. The purified protein had a molecular mass of 14,800 Da and stimulated the proliferation of capillary endothelial cells. Also, it stimulated new vessel growth in vivo in the chick CAM assay (Shing et al., 1985). The molecule was the very same agent that Gospodarowicz and his co-workers had identified years earlier but had not purified.

Heparin affinity chromatography allowed the purification of FGF from different sources and the identification of two prototypic members of the FGF family [acidic FGF (aFGF), eluting from a heparin–Sepharose column with a 1.0 M NaCl wash, and basic FGF (bFGF), eluting with a 1.5 M NaCl wash] (Gospodarowicz et al., 1984; Esch et al., 1985). In 1986, an angiogenesis factor was isolated from human placenta and human hepatoma cells on the basis of its ability to stimulate protease production in cultured capillary endothelial cells (Moscatelli et al., 1986a; Presta et al., 1986). The purified factor also stimulated DNA synthesis and motility in capillary endothelial cells and induced angiogenesis in vivo. Amino acid sequence data revealed that the angiogenesis factor was human bFGF. Since then, numerous studies have shown that FGF-2 is a pleiotropic factor modulating cell proliferation and/or differentiation in a variety of cell types, thus affecting various organ sys-tems and biological processes, including neoplasia (Baird et al., 1986). Today, FGFs comprise a structurally related family of 22 molecules (Itoh and Ornitz, 2004). On the basis of the new nomenclature, FGFs are numbered consecutively, bFGF being named FGF-2.

2.2.1 FGF-2 Receptors and FGF-2 Interactions

To exert their biological activity, FGFs interact with high-affinity tyrosine kinase FGF receptors (FGFRs). Four members of the FGFR family (FGFR-1, FGFR-2, FGFR-3, and FGFR-4) are encoded by distinct genes and their structural variability is increased by alternative splicing (Johnson and Williams, 1993). The various FGFs show different FGFR specificity, FGF-1 being the universal FGF that can activate all FGFRs. FGF-2 binds to distinct splice variants of the different FGFRs, FGFR-1c isoform being the preferential target of FGF-2 (Zhang et al., 2006). FGFR-1 is expressed by endothelial cells in vivo and in vitro. Less frequently, cultured en-dothelial cells can express FGFR-2 (Dell'Era et al., 2001), whereas the expression of FGFR-3 or FGFR-4 has never been reported in endothelium. FGFRs are expressed on nearly every cell type of hematopoietic origin and deregulation of FGFR gene expression and/or gene mutation has been observed in hematologic malignancies (Moroni et al., 2002).

In endothelium, activation of FGFR-1 or FGFR-2 by FGF-2 leads to endothelial cell proliferation (Cross and Claesson-Welsch 2001) and to the induction of a complex pro-angiogenic phenotype. However, various other binding partners can affect the biological activity and angiogenic potential of FGF-2. These molecules can interact with FGF-2 in the extracellular environment, thus modulating its bioavailability, stability, local concentration, interaction with endothelial receptors, and intracellular fate. They include heparan-sulfate proteoglycans (HSPGs), integrin receptors, extracellular matrix proteins, serum components, and cytokines (Presta et al., 2005).

2.2.2 *The Role of FGF-2 in Tumor Growth and Vascularization*

A large body of research has implicated the FGF/FGFR system as having a role in tumorigenesis (Gross and Dickson, 2005). Mouse models have confirmed that FGFs, including FGF-2, may exert an oncogenic effect. Since no activating mutations have been detected in FGFs themselves, the simplest mechanism by which FGF-2 may contribute to tumor cell proliferation is by overexpression and release of themselves from tumor cells and/or from stromal cells. Also, dysregulation of FGF-2 signaling as a result of alterations at the FGFR level may play a role in cancer. These alterations include inappropriate expression, activating point mutations, splice variations, and genomic alterations.

Various tumor cell lines express FGF-2 (Presta et al., 1986; Moscatelli et al., 1986b) and the appearance of an angiogenic phenotype correlates with the export of FGF-2 during the development of fibrosarcoma in a transgenic mouse model (Kandel et al., 1991). Also, FGF-2 can be expressed by activated endothelium. Gualandris et al. (1996) originated a stable mouse aortic endothelial cell line transfected with a human FGF-2 cDNA. Endothelial FGF-2 transfectants show an invasive and morphogenetic behavior in vitro. In vivo, they are angiogenic, cause the formation of opportunistic vascular tumors in nude mice, and induce hemangiomas in the chick embryo (Ribatti et al., 1999b). FGF-2 transfection affects the expression of numerous genes implicated in the modulation of cell cycle, differentiation, cell adhesion, and stress/survival (Dell'Era et al., 2002). Some of these genes are similarly modulated in vitro and in vivo by administration of the recombinant growth factor (Dell'Era et al., 2002).

Early studies showed that elevated levels of FGF-2 in urine samples collected from 950 patients having a wide variety of solid tumors, leukemia, or lymphoma were significantly correlated with the status and the extent of the disease (Nguyen et al., 1994). However, no association between increased serum levels of FGF-2 and tumor type was observed in later studies on a large spectrum of metastatic carcinomas even though two-thirds of the patients showing progressive disease had increasing serum levels of the angiogenic factor compared with less than one-tenth of the patients showing response to therapy (Dirix et al., 1997). Also, serum concentration of FGF-2 has prognostic relevance for advanced head and neck cancer

(Dietz et al., 2000) even though serum FGF-2 may not entirely be derived from the neoplastic tissue in cancer patients (Salgado et al., 2004).

After an encouraging report about a positive correlation between microvessel density and cerebrospinal fluid FGF-2 in children with brain tumors (Li et al., 1994), it is seen that FGF-2 levels in body fluids do not always reflect tumor vascularity. Indeed, numerous studies have attempted to establish a correlation between intra-tumoral levels of FGF-2 mRNA or protein and intratumoral microvessel density in cancer patients. In fact, because of its pleiotropic activity that may affect both tumor vasculature and tumor parenchyma, FGF-2 may contribute to cancer progression not only by inducing neovascularization, but also by acting directly on tumor cells. Indeed, cancer epithelial cells from pancreatic (Ohta et al., 1995), breast (Yiangou et al., 1997), non-small cell lung (Volm et al., 1997), and head and neck squamous carcinomas (Dellacono et al., 1997) show an increased production of FGF-2 by the tumor cell themselves. On the other hand, FGFRs are overexpressed and/or mutated in several human cancers (Birnbaum et al., 1991). The capacity of tumor, stromal, and endothelial cells to express both FGF-2 and its receptors points to autocrine and paracrine functions of this growth factor in different cancers, including hematopoi-etic neoplasm (Ribatti et al., 2007). FGFs are potent activators of endothelial pro-liferation and can thus stimulate angiogenesis, promote stromal fibroblast prolifer-ation, and extracellular matrix formation leading to excessive bone marrow fibrosis and can directly affect neoplastic cells by acting on their high-affinity FGFRs.

2.3 The Discovery of Vascular Permeability Factor/Vascular Endothelial Growth Factor

2.3.1 *The Contribution of Harold F. Dvorak (Fig. 2.2). Tumor Blood Vessels Are Hyperpermeable to Plasma Proteins and to Other Circulating Macromolecules*

Back in the 1970s, Dvorak investigated the cellular composition of delayed-type hypersensitivity reactions to soluble protein antigen in guinea pigs and discovered that basophilic leukocytes were a prominent component (Dvorak et al., 1970). Then, Dvorak used as antigens two guinea pig tumor cell lines and demonstrated that the immune response elicited by these tumors included basophils and macrophages (Dvorak et al., 1973). Moreover, within days of the transplant, the tumors were organized into clumps of cells that were separated by spaces containing thin strands of fibrillary material, constituted by cross-linked fibrin, as demonstrated by electron microscopy, immunohistochemistry, and biochemistry (Dvorak et al., 1979a,b).

Once deposited, cross-linked fibrin behaves as a gel that causes edema by trafficking extravasated plasma and provides a pro-angiogenic stroma. In fact, en-dothelial cells, fibroblasts, and inflammatory cells synthesize and secrete the ma-trix proteins, proteoglycans, and glycosaminoglycans that comprise mature tumor stroma, and express adhesion molecules whose interaction with fibrin allows them

Fig. 2.2 A portrait of Harold F. Dvorak

to move freely in tumor stroma. Finally, the fibrin matrix supports proliferation of the tumor cells.

Dvorak demonstrated that vascular hyperpermeability to fibrinogen and other plasma proteins, as well as fibrin deposition, is a common feature of many animal and human tumors, both transplantable and autochthonous (Brown et al., 1988; Dvorak et al., 1981; 1983; 1984; Harris et al., 1982). Hyperpermeable vessels were especially prominent at the tumor–host interface and it was therefore not certain whether tumor cells were permeabilizing normal host microvessels and/or were generating the formation of new, abnormal blood vessels that were intrinsically permeable.

2.3.1.1 Vascular Permeability Factor Activity Is Present in Tumor Culture Supernatants

Testing cell-free supernatants from a variety of human and animal tumor cells using Miles assay (Miles and Miles, 1952), Dvorak found that supernatants from nearly all of them generated an intense blue spot due to extravasated Evans blue, whereas those from several normal cells did not (Dvorak et al., 1979b). Dvorak called this tumor supernatant permeabilizing activity as vascular permeability factor (VPF). With a potency some 50,000 times that of histamine (Dvorak et al., 1992; Senger et al., 1983) VPF was effective at concentrations well below 1 nM in the Miles assay.

Dvorak showed that VPF was non-dialyzable and therefore likely to be a macromolecule, while inhibition of protein synthesis profoundly depressed its secretion and heat and proteases largely inactivated its activity (Senger et al., 1983). VPF does not itself provoke mast cell degranulation or induce a significant inflammatory cell infiltrate. The permeabilizing action of VPF was not blocked by inhibitors of inflammation, including those that block histamine, thrombin, and platelet activating factor (Senger et al., 1993).

VPF increases the permeability of microvessels, primarily post-capillary venules and small veins, to circulating macromolecules. VPF permeabilized a number of vascular beds, including those of the skin, subcutaneous tissues, peritoneal wall, mesentery, and diaphragm (Dvorak et al., 1979b; Collins et al., 1993; Nagy et al., 1995).

2.3.1.2 The Discovery of Vascular Permeability Factor

Senger purified VPF to homogeneity with heparin–Sepharose and hydroxylapatite chromatography, and demonstrated that VPF is a 34–43-kDa dimeric protein whose activity was lost by reduction, but was unaffected by deglycosylation (Senger et al., 1983). However, the affinity of VPF for heparin was substantially lower than that of other typical heparin-binding growth factors, such as basic fibroblast growth factor (Senger et al., 1983). Senger sequenced the N-terminus and made use of this sequence to prepare a rabbit antibody against a peptide corresponding to the first 24 amino acids of VPF (Senger et al., 1983). This antibody abolished all of the permeability-increasing activity present in culture medium from several guinea pig and rat tumors, and prevented circulating albumin from accumulating in tumor ascites fluid.

Subsequently, Connolly and co-workers at Monsanto Company showed that VPF is an endothelial mitogen in vitro and an angiogenic factor in vivo (Connolly et al., 1989).

At low levels VPF does not increase vascular permeability, or induce angiogenesis and it might have other functions in normal physiology such as acting as an endothelial cell survival factor (Benjamin and Keshet, 1997) or preventing endothelial cell apoptosis and senescence (Watanabe et al., 1997; Benjamin et al., 1999).

Fluid accumulation results from VPF-induced leakage of plasma through hyperpermeable microvessels, but is also favored by the fact that tumors in general lack lymphatic vessels and hence are unable to drain extravasated proteinaceous fluid effectively.

2.3.1.3 Tumors: Wounds That Not Heal

Dvorak pointed out that similarities exist between tumor stroma generation and wound healing. He noted that wounds, like tumors, secrete VPF, causing blood vessels to leak plasma fibrinogen, which stimulates blood vessel growth and provides a matrix on which they can spread. Unlike wounds, however, that turn off VPF production after healing, tumors did not turn off their VPF production and instead continued to make large amounts of VPF, allowing malignant cells to continue to induce new blood vessels and so to grow and spread. Thus, tumors behave as wounds that fail to heal (Dvorak, 1986).

Wounds in rodent skin, like tumors, secrete VPF: within 24 h of wounding VPF mRNA expression increases in epidermal keratinocytes at the wound edge (Brown et al., 1992). VPF overexpression reaches a peak at 2–3 days and persists at an elevated level for about 1 week, the time required for granulation tissue to form and migrating keratinocytes to cover the wound defect. In contrast to tumors, VPF expression was downregulated as healing progressed and, parallel with the decreased expression of VPF, vascular permeability returned to normal.

In contrast to normal mice, congenitally diabetic *db/db* mice have elevated endogenous levels of VPF mRNA in their nude skin, which increase transiently after wounding. However, the rise of VPF is not sustained and as granulation tissue forms, VPF expression plummets to barely detectable levels, thus associating decreasing VPF expression with defective wound healing (Peters et al., 1993).

2.3.2 The Contribution of Napoleone Ferrara (Fig. 2.3)

In 1989, Ferrara and Henzel reported the isolation of a diffusible endothelial cell-specific mitogen from a medium conditioned by bovine pituitary follicular cells, which they named "vascular endothelial growth factor" (VEGF) to reflect the restricted target cell specificity of this molecule. NH_2-terminal amino acid sequencing of purified VEGF proved that this protein was distinct from the known endothelial cell mitogens such as aFGF and bFGF and indeed did not match any known protein in available databases (Ferrara and Henzel, 1989). By the end of 1989, Ferrara reported the isolation of cDNA clones for bovine VEGF164 and three human VEGF isoforms: VEGF 121, VEGF 165, and VEGF 189 (Leung et al., 1989). Subsequent studies indicated that these isoforms had markedly different properties in terms of diffusibility and binding to heparin. VEGF 121, which lacked heparin-binding, was highly diffusible, whereas VEGF 189, a highly basic and heparin-binding protein, was almost completely sequestered in the extracellular matrix and VEGF 165 had intermediate properties (Houck et al., 1991). Additionally, some proteases like plasmin were found to cleave heparin-binding VEGF isoforms in the COOH terminus and thus generating a non-heparin-binding diffusible fragment (Houck et al., 1992). These early studies suggested that both alternative RNA splicing and extracellular proteolysis regulate the activity of VEGF.

Over the years, five VEGF-related genes have been identified (VEGF-A, VEGF-B, VEGF-C, VEGF-D, and VEGF-E). There are five characterized VEGF-A isoforms of 121, 145, 165, 189, and 206 amino acids in mammals, generated by alternative splicing of the mRNA from a single gene comprising eight exons. They display differential interactions with related receptor tyrosine kinases VEGFR-1/Flt-1, VEGFR-2/Flk-1, VEGFR-3/Flt-4, and neuropilin-1 and neuropilin-2 (NRP-1 and

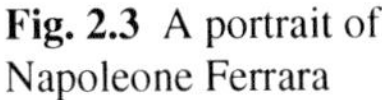

Fig. 2.3 A portrait of
Napoleone Ferrara

NRP-2). As a result of the receptor activation and subsequent signal transduction,
VEGF target cells may proliferate, migrate, or alter gene expression, e.g., of MMPs
or cytokines.

VEGFR-1 and VEGFR-2 are restricted largely to vascular endothelium in their
expression, accounting for the specificity of action of this growth factor family. In
1992 in a collaborative study between Ferrara's laboratory and Lewis Williams's
group at the University of California at San Francisco, VEGFR-1 was shown to
be a high-affinity VEGF receptor (de Vries et al., 1992). Ferrara also demonstrated that VEGFR-1 expression is upregulated by hypoxia via a hypoxia inducible
factor (HIF)-1-dependent mechanism (Gerber et al., 1997) and that VEGFR-1
binds not only VEGF-A, but also placental growth factor (PlGF) (Park et al.,
1994).

Ferrara initially proposed that VEGFR-1 may be not primarily a receptor trans-
mitting a mitogenic signal, but rather a "decoy" receptor, able to regulate in a
negative fashion the activity of VEGF on the vascular endothelium, by sequester-
ing and rendering this factor less available to VEGFR-2 (Park et al., 1994). Thus,
the observed potentiation of the action of VEGF by PlGF could be explained, at
least in part, by the displacement of VEGF from VEGFR-1 binding (Park et al.,
1994).

VEGFR-3 is restricted largely to the lymphatic endothelium (Kukk et al., 1996). VEGFR-3 may play a role in disorders involving the lymphatic system and angiogenesis and may be of potential use in drug targeting, in vivo imaging of the lymphatic vessels, and in therapeutic lymphangiogenesis. VEGF-C binds to VEGFR-3, expressed on lymphatic endothelium, and has been implicated in lymphangiogenesis. Like VEGF-C, to which is structurally related, VEGF-D is an endothelial cell mitogen and interacts with VEGFR-2 and VEGFR-3. VEGF-E, encoded by the ORF virus, induces angiogenesis through an interaction with VEGFR-2 (Meyer et al., 1999). Overexpression of VEGF-C and VEGF-D in transgenic mice induces the formation of hyperplastic lymphatic vessels. Conversely, inhibition of VEGF-C and/or VEGF-D by overexpression of a soluble form of VEGFR-3 in the skin of transgenic mice leads to inhibition of lymphatic vessel growth (Jussila and Alitalo, 2002). Transgenic inactivation of both VEGF-C alleles results in prenatal death: endothelial cells commit to the lymphatic lineage, but do not sprout from veins (Karkkainen et al., 2004).

NRP-1 is important for both blood vessel development and development of the nervous system and is a receptor for semaphorin-3A which acts as an axonal-repellent factor. NRP-1 forms complexes with either VEGFR-1 or VEGFR-2 and is an enhancer of VEGFR-2 activity (Soker et al., 1998; Whitaker et al., 2001). In this way, NRP-1 contributes to the sum of pro-angiogenic functions mediated by VEGFR-2, and it might also participate in endothelial guidance and vascular patterning (Gerhardt et al., 2004). Moreover, stimulation by nerve growth factor (NGF) and VEGF activates two common intracellular signaling cascades in endothelial cells, the Ras/ERK and P13K/Akt pathways, both of which are involved in cell proliferation and survival, suggesting that NGF, acting in concert with VEGF, plays a role in controlling angiogenic processes (Nico et al., 2008b).

2.3.2.1 Role of VEGF in Embryonic Vasculogenesis and Angiogenesis

In 1996, Ferrara's laboratory (Ferrara et al., 1996) and a collaborative effort between Peter Carmeliet in Leuven, Werner Risau in Martinsried, and Andras Nagy in Toronto (Carmeliet et al., 1996) demonstrated an essential role of VEGF in embryonic vasculogenesis and angiogenesis in the mouse. Inactivation of a single VEGF allele resulted in embryonic lethality between day 11 and day 12. The VEGF+/− embryos exhibited a number of developmental anomalies. The forebrain region appeared significantly underdeveloped. In the heart, the outflow region was grossly malformed; the dorsal aortas were rudimentary, and the thickness of the ventricular wall was markedly decreased. The yolk sac revealed a substantially reduced number of nucleated red blood cells within the blood islands, indicating that VEGF regulated both vasculogenesis and hematopoiesis. Also, the vitelline veins failed to fuse within the vascular plexus of the yolk sac. Significant defects in the vasculature of other tissues, including placenta and nervous system, were evidenced. For example, in the nervous system of heterozygous embryos at day 10.5, vascular elements could be demonstrated in the mesenchyme, but not in neuroepithelium and the failure of blood vessels in growth was accompanied by apoptosis and disorganization of neuroepithelial cells.

2.3.3 *The Discovery of Angiopoietins*

A novel family of angiogenic factors, designated as angiopoietins (Ang), has been identified by Yancopoulous and co-workers (Davis et al., 1996; Maisonpierre et al., 1997). Ang-1 was discovered originally as a ligand for Tie-2, a member, with the originally cloned isoform Tie-1, of the tyrosine kinase with immunoglobulin and epidermal growth factor homology receptor (Tie) family (Davis et al., 1996; Dumont et al., 1992; Partanen et al., 1992; Sato et al., 1993). Ang-2 also binds Tie-2 (Maisonpierre et al., 1997). Although, both Ang-1 and Ang-2 bind Tie-2, no ligand for Tie-1 has been identified. Ang plays a role in vascular stabilization (Beck and D'Amore, 1997). Ang-1 is associated with developing vessels and its absence leads to defect in vessel remodeling. Ang-2, which antagonizes the action of Ang-1, plays a role in the destabilization of existing vessels (it is found in tissues like the ovary, uterus, and placenta that undergo transient or periodic growth and vascularization, followed by regression). In the absence of Ang-1, angiogenic factors like VEGF may produce immature vessels that are hemorrhagic and display poor contact with underlying matrix material. Tie-2 (Sato et al., 1995) or Ang-1 (Suri et al., 1996) knock-out mice show immature vascularization pattern as well as lack of periendothelial mesenchymal cells, such as pericytes and immature smooth muscle cells (SMCs), leading to death around 11.0–12.5 days of gestation. Targeted disruption of the Tie-1 gene indicates that it is required for the maintenance of vascular integrity (Sato et al., 1995). Ang-2 knock-out mice display a generalized lymphatic dysfunction caused by disorganized collecting lymphatic vessels with poorly associated vascular SMCs and an irregularly patterned hypoplastic lymphatic capillary network (Gale et al., 2002). Transgenic mice overexpressing Ang-2 died during embryogenesis with similar vascular defects as mice lacking Ang-1 or Tie-2 (Maisonpierre et al., 1997).

Ang-2 seems to be the earliest marker of blood vessels that has been perturbed by invading tumor cells (Holash et al., 1999b). Ang-2 is overexpressed in tumor microvasculature of human glioblastoma and hepatocellular carcinoma (Stratmann et al., 1998; Tanaka et al., 1999) and is upregulated together to Ang-1 in several types of tumor (Table 2.1).

Table 2.1 Human tumors in which Ang-1 and Ang-2 are upregulated

Stomach
Breast
Colon
Liver
Brain
Lung
Leukemia

Chapter 3

3.1 How Do Tumor Cells Switch to the Angiogenic Phenotype?

Spontaneously arising tumor cells are not usually angiogenic at first (Folkman et al., 1989). The phenotypic switch to angiogenesis is usually accomplished by a subset that induces new capillaries which then converge toward the tumor. These new vessels perfuse the tumor with blood, which transports nutrients and oxygen to the tumor and catabolites away from it, and their endothelial cells produce a spectrum of growth factors with a paracrine stimulatory effect on the tumor cells and a variety of matrix-degrading proteinases that facilitate invasion (Nicosia et al., 1986). An expanding endothelial surface also gives tumor cells more opportunities to enter the circulation and metastasize, while their release of antiangiogenic factors explains the control exerted by primary tumors over metastasis. These observations suggest that tumor angiogenesis is linked to a switch in the equilibrium between positive and negative regulators. In normal tissues, vascular quiescence is maintained by the dominant influence of endogenous angiogenesis inhibitors over angiogenic stimuli. Tumor angiogenesis, on the other hand, is induced by the increased secretion of angiogenic factors and/or downregulation of such inhibitors.

Neovascular channels allow tumor cells to metastasize hematogenously. For example, 10–100 million endothelial cells are required to support the smallest palpable breast cancer. It is 1 cm in diameter, weighs 1 g, and consists of approximately 1 billion cancer cells (Modzelewski et al., 1994). It sheds 2×10^6 cancer cells into the systemic circulation every 24 h (Butler and Gullino, 1975). Of these, only one survives. It lodges in a distant organ as a micrometastasis and remains quiescent until the angiogenesis occurs in its site.

Most human tumors arise and remain in situ without angiogenesis for months to years before switching to an angiogenic phenotype, though the preneoplastic stage of breast and cervical carcinomas becomes neovascularized before the malignant tumor appears. In human primary tumors, microscopic areas of intense angiogenesis flank less vascularized areas (Weidner et al., 1991), suggesting a heterogeneity of clones of highly angiogenic cells as well as weakly angiogenic or non-angiogenic clones. In cervical neoplasia, a switch is readily apparent in mid-late dysplasias, wherein new vessels became densely apposed along the basement membrane underlying the dysplastic epithelium (Guidi et al., 1995).

D. Ribatti, *History of Research on Tumor Angiogenesis*,
DOI 10.1007/978-1-4020-9563-4_3, © Springer Science+Business Media B.V. 2009

3.2 The Concept of Angiogenic Switch

The angiogenic switch whereby the normally quiescent vasculature grows new capillaries separates the avascular (prevascular) phase characterized by a dormant tumor and the vascular phase in which exponential tumor growth ensues (Fig. 3.1). In the prevascular phase, tumor cells proliferate (sometimes as rapidly as in the vascularized tumor), but the rate of tumor cell death (apoptosis) counterbalances this proliferation and maintains the tumor mass in a steady state. Dormant tumors have been discovered during autopsies of individuals who died of causes other than cancer (Kirsch et al., 2004). This supports the notion that only a very small subset of dormant tumors enter the second phase, the vascular phase in which exponential tumor growth ensues.

Activation of the switch itself has been attributed to the synthesis or release of angiogenic factors. The balance hypothesis (Hanahan and Folkman, 1996; Bouck et al., 1996) assumes that the level of angiogenesis inducers and inhibitors governs cell differentiation states of quiescence or angiogenesis (Fig. 3.2). This balance is altered by increasing activator gene expression, changing the bioavailability or activity of the inducer proteins, or reducing the concentrations of endogenous angiogenesis inhibitors, here, too, via changes in gene expression or processing/biovailability.

In 2001, Achilles and co-workers reported that human tumors contain subpopulations that differ in their angiogenic potential. They established and selected subclones from a human liposarcoma cell line (SW-872) based on high, intermediate, or low proliferation rates in vitro. These clones were subsequently expanded in vitro into distinct populations of tumor cells and were then inoculated into

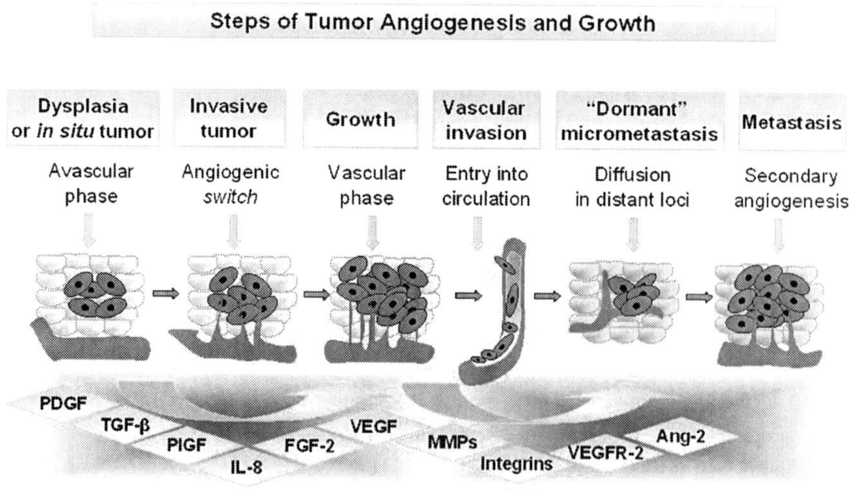

Fig. 3.1 Steps of tumor angiogenesis and growth. (Reproduced from Ribatti D, Vacca A, Overview of angiogenesis during tumor progression. In "Angiogenesis. An Integrative Approach from Science to Medicine", Figg WJ, Folkman J, eds., Springer Science, New York, 2008, pp. 161–8.)

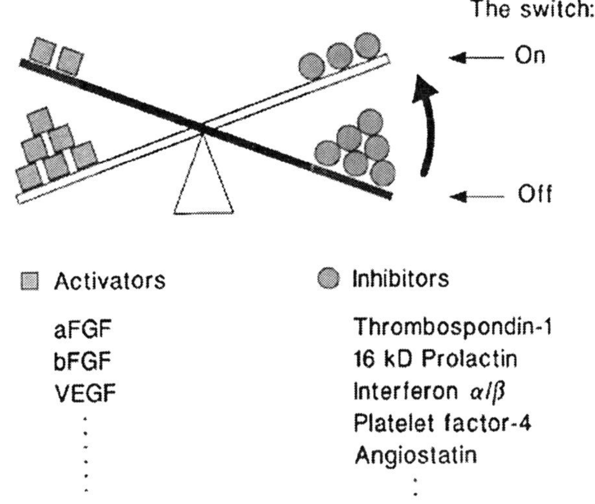

Fig. 3.2 The balance hypothesis for the angiogenic switch. (Reproduced from Hanahan D, Folkman J, Cell, 86: 356–64, 1996.)

immunodeficient [severe combined immuno deficiency (SCID)] mice. Three different in vivo growth patterns were observed: (i) high angiogenic and rapidly growing tumors; (ii) weakly angiogenic and slowly growing tumors; (iii) non-angiogenic and dormant tumors.

Further investigation by Almong et al. (2006) showed that the non-angiogenic tumors spontaneously switch to the angiogenic phenotype and initiate exponential growth approximately 130 days after inoculation into the subcutaneous space. During the 130 days dormancy period, microscopic tumors remained avascular and were virtually undetectable by palpation.

3.3 Factors Involved in the Angiogenic Switch

The switch depends on increased production of one or more of the positive regulators of angiogenesis, such as VEGF, FGF-2, IL-8, transforming growth factor beta (TGF-β), platelet derived growth factor (PDGF), pleiotrophins, and others. These can be exported from tumor cells (Kandel et al., 1991), mobilized from the extracellular matrix (Vlodavski et al., 1990), or released from host cells (e.g., macrophages) recruited to the tumor (Leibovich et al., 1987). Expression of endogenous inhibitors, such as thrombospondin-1 (TSP-1) or interferon beta may be downregulated (Rastenejad et al., 1989; Good et al., 1990; Bornstein, 1992; Dameron et al., 1994). Thus, the switch clearly involves more than simple upregulation of angiogenic activity and has thus been seen as the result of a net balance of positive and negative regulators.

Integrin signaling also contributes to this regulatory balance. Quiescent vessels express one class of integrins, whereas sprouting capillaries express another. Interference with signaling by the latter class of integrins can inhibit angiogenesis (Giancotti and Ruoslahti, 1999). In vivo screening of phage libraries in murine models has identified specific motifs, including RGD, GSL, and NGR, that bind to integrins $\alpha v\beta 3$, $\alpha v\beta 5$ and $\alpha 5\beta 1$, MMPs, and VEGFR that are upregulated in neoangiogenic tumor endothelial cells (Pasqualini and Arap, 2002; Ruoslahti, 2002).

Proteases control the bioavailability of angiogenic activators and inhibitors. Some release FGF-2 stored in the extracellular matrix (Whitelok et al., 1996), whereas plasmin, a pro-angiogenic component of the clotting system, cleaves itself into an angiogenesis inhibitor form, namely angiostatin (Gately et al., 1997).

Nutrient deprivation modulates gene expression and may also contribute to the activation of the angiogenic process. Glucose deprivation-induced oxidative stress activates the expression or release of angiogenic growth factors (Spitz, 2000).

3.4 What Is the Evidence that Genetic Instability Promotes the Angiogenic Switch?

A widely accepted view is that the progression of tumors reflects their genetic instability, this being defined as their higher mutation rate compared with normal tissues as corruptions in checkpoint genes are crucial for genome replication (Lengauer et al., 1998). Progression is thus achieved through the accumulation of multiple lesions that impair the control of cell proliferation and survival and thereby shape the complex phenotype of tumor cells (Hanahan and Weinberg, 2000). Genetic instability may be required for the emergence of angiogenic tumor cell lines that enhance a tumor's growth and malignancy. In the absence of such instability, these lines cannot grow, even if the relevant mutations are generated at low levels because the angiogenic promoters will not be sufficient to counter the influence of inhibitory factors. Inhibition of angiogenesis can thus be viewed as a host defense and a tumor must be genetically unstable to be able to exceed a certain size. Genetic control of the physiological levels of endogenous angiogenesis inhibitors may well be a line of defense against the conversion of dormant tumor cells to a malignant phenotype.

Genetic instability must therefore act upstream and promote the angiogenic switch. Evidence in favor of this view has been recently acquired from reversible transgene models and multigene-transformed cells. Watnick et al. (2003) observed that the switch in a cell transformation model was dependent on oncogenic *RAS* expression. They showed that low expression levels induced cell transformation and increased VEGF expression, and that further increases in the abundance of the oncogene led to repression of the antiangiogenic factor TSP-1 through *Myc* activation, and subsequent tumor expansion. Expression of the *SV40* early region, *TERT*, and activated *RAS* was sufficient to transform primary epithelial cells in vitro (Dameron et al., 1994). However, their ability to grow in vivo depended on the level of *HRAS* expression: cells that expressed low levels were dormant and non-angiogenic, whereas those that expressed high levels progressed to full-blown tumors. CEGF-A levels increased only modestly (1.4-fold) in tumors expressing high-*RAS* levels, whereas TSP-1 levels increased 8-fold. Tumor formation from cells that expressed low RAS levels was indicated by simple overexpression of VEGF-A. This shows tumor progression was blocked because the angiogenic switch was not activated. It also supports the view that the switch is determined by the balance between pro- and antiangiogenic factors, and that oncogene expression can influence this balance. Analysis of 15 of the most studied oncogenes revealed that the majority of them increase the expression of VEGF and/or FGF-2 and decrease the expression of TSP-1 in tumor cells (Rak et al., 2002).

BHK 21/cl 13 cells, an immortal but non-tumorigenic line of hamster fibroblasts, were converted to malignancy and anchorage independence by loss of a functioning tumor suppressor gene. These cells were highly tumorigenic in nude mice and neonatal hamsters, and potently angiogenic in vivo. Normal BHK cells and suppressed hybrids generated by fusing transformed BHK cells with either non-transformed BHK or normal human fibroblasts were unable to induce angiogenesis when they or their concentrated conditioned media were introduced into rat corneas, whereas transformed BHK cells and transformed segregants from the suppressed hybrids were angiogenic. Mixing experiments showed that normal cells elaborated an angiogenesis inhibitor whose production was blocked coincidentally with suppressor loss. When endothelial cell chemotaxis was used as an in vitro corollary of angiogenesis in the rat cornea assay, the inhibitor was purified and shown to be TSP-1 (Bornstein, 1992). This was the first illustration of a new function for a tumor suppressor gene, namely regulation of the production of a naturally occurring inhibitor of angiogenesis. In another set of experiments, Dameron et al. (1994) established a direct link between the *p53* tumor suppressor gene, tumor angiogenesis, and TSP-1. To examine the effect of *p53*, they used cultured fibroblasts from patients with the Li-Fraumeni syndrome, who have inherited one wild-type allele and one mutant allele of the *p53* gene. When the wild-type allele was lost, these cells acquired potent angiogenic activity coincidental with the loss of TSP-1 production. Transfection revealed that *p53* stimulated the endogenous TSP-1 gene and positively regulated the TSP-1 promoter sequences.

3.4.1 The RIP1-TAG2 Model

The mechanism of the switch was first described in 1985 by Hanahan who developed transgenic mice in which the large *T* oncogene is hybridized to the insulin promoter. In this islet cell tumorigenesis (*RIP1-TAG2* model), these mice express the large T antigen in all their islet cells at birth, and express the SV40 T antigen (*TAG*) under the control of the insulin gene promoter, which elicits the sequential development of tumors in the islets over a period of 12–14 weeks. Tumor development proceeds by stages during which about half of the 400 islets become hyperproliferated, while a subset (about 25%) subsequently acquire the ability to switch to angiogenesis (Folkman et al., 1989). Some 15–20% of these angiogenic islets develop into benign tumors, encapsulated lesions, and invasive carcinomas (Lopez and Hanahan, 2002). This multistage pathway suggests the sequential involvement of multiple rate-limiting genetic and epigenetic events in the progression from normal cells to tumors. The β cells become hyperplastic and progress to tumors via a reproducible and predictable multistep process (Hanahan, 1985). One step occurs at 6–7 weeks, when angiogenesis is switched on in approximately 10% of preneoplastic islets. Solid vascularized tumors first appear at 9–10 weeks, initially as small nodules that grow and progress to large islet tumors, with well-defined margins, as well as two classes of invasive carcinoma (Lopez and Hanahan, 2002). Lopez and Hanahan identified stage-specific molecular markers accessible via the circulation, either on the surface of endothelial cells, their peri-endothelial support cells (pericytes and SMCs), or even tumor cells themselves (as a result of the hemorrhagic leaky angiogenic vasculature). They selected phage pools that homed preferentially to different stages during *RIP1-TAG2* tumorigenesis. In addition to "panangiogenic" markers shared by many types of tumors, they identified vascular target molecules characteristic of this tumor's tissue of origin and not expressed in the vessels of several tumor types growing in or under the skin.

Angiogenic islets are revealed both morphologically in tissue sections and in isolated islets by their red color and microhemorrhagic islands, and functionally by their ability to elicit endothelial cell migration, proliferation, and tube formation in an in vitro collagen bioassay involving coculture of dispersed capillary endothelial cells and isolated islets (Folkman et al., 1989). This onset pattern closely resembles that of angiogenesis in human tumors.

Two concepts emerged from this early characterization of tumorigenesis in *RIP-TAG* transgenic mice: (i) the existence of distinct stages of premalignant progression, namely a hyperplastic stage followed by a stochastic angiogenic stage and (ii) the development of angiogenesis well before the emergence of an invasive malignancy. The temporal and histological changes that occur in the *RIP-TAG* model are consistent with the multistep paradigm for tumorigenesis of human cancers (Vogelstein, 1993). The high incidence of occult human cancers suggests that this angiogenic switch may, as in the *RIP-TAG* model, be a relatively late event that plays a significant role in the transition from microscopic foci to macroscopic tumor (Udagawa et al., 2002).

These data suggested that the induction of angiogenesis during multistage carcinogenesis is coordinated by an angiogenic switch. VEGF signaling is primarily implicated in angiogenesis and tumorigenesis in *RIP1-TAG2* mice. The islets ar extensively vascularized to facilitate their monitoring of serum glucose levels and hence the secretion of insulin and other hormones for endocrine regulation of carbohydrate metabolism. VEGF-A, VEGF-B, and VEGF-C are all expressed in normal islet β cells (Christofori et al., 1995). VEGF-A is expressed at all stages of *RIP1-TAG2* tumorigenesis (Christofori et al., 1995). Such constancy suggests that if VEGF-A activity is important in this tumorigenesis pathway, other modes of regulation may be involved. Inoue et al. (2002) showed that five VEGF ligand genes are expressed in normal islets and throughout tumorigenesis. Moreover, they produced a β cell-specific VEGF-A knockout that resulted in islets with reduced vascularity, but essentially normal physiology. In *RIP1-TAG2* mice where most oncogene-expressing cells had deleted the VEGF-A gene, both angiogenic switching and tumor growth were severely disrupted, as was the neovasculature.

Overexpression of VEGF-C in β cells via a *RIP-VEGF-C* transgene induces peri-islet lymphatic vessels, but has no discernable effect on intra-islet blood vessels (Mandriota et al., 2001). Extensive lymphatic channel formation, in fact, was observed in around (but not within) 98% of islets from *RIP-VEGF-C* transgenic mice, i.e., the anatomical units in which the transgene is expressed. VEGF-C overexpression did not increase intratumoral vascularity or enhance tumor growth, though in peri-tumoral lymphatic vessels it induced facilitated metastases to the draining mesenteric lymph nodes. To assess the role of VEGF-C-induced lymphangiogenesis in tumor metastasis, Mandriota et al. (2001) crossed *RIP-VEGF-C* mice with *RIP1-TAG2* mice. β-cell tumors in *RIP1-TAG2* mice invade locally, but are not metastatic. VEGF-C expression in double transgenics resulted in the de novo formation of lymphatics in intimate association with β-cell tumors associated with the formation of metastases in the draining regional mesenteric lymph nodes in 37% of mice.

Bergers et al. (2000) demonstrated that MMP-9 plays a crucial role in the initial angiogenic switch during islet carcinogenesis and proposal mobilization of VEGF from an extracellular reservoir as its mode of action. Moreover, preclinical trials targeting MMP-9 and angiogenesis with an MMP inhibitor and with a bisphosphonate, zoledronic acid, showed that both were antiangiogenic. Other enzymes are involved in islet tumorigenesis. Joyce et al. (2004) have shown that a subset of papain family Clan CA proteases known as cathepsins make an important contribution to the development of islet tumors and are upregulated during their progression. Cathepsin activity was assessed with chemical probes to allow biochemical and in vivo imaging. Increased activity was associated with the angiogenic vasculature and invasive fronts of carcinomas, with differential expression in immune, endothelial, and cancer cells. A broad-spectrum cysteine inhibitor that knocked out cathepsin function at different stages of tumorigenesis impaired angiogenic switching in progenitor lesions, as well as tumor growth, vascularity, and invasiveness. Cysteine cathepsins are also upregulated during *HPV16*-induced cervical carcinomas. Joyce et al. (2005) have since shown that heparanase expression increases during *RIP-TAG* tumorigenesis, predominantly supplied by innate immune cell infiltrating neoplastic tissues.

Joyce et al. (2003) analyzed the vasculature in the angiogenic stages of *RIP-TAG* model islet tumorigenesis with phage libraries that display short peptides, and identified peptides that discriminate between the vasculature of the premalignant angiogenic islets and the fully developed vasculature. One peptide is homologous with PDGF-B, which is expressed in endothelial cells, while its receptor is expressed in pericytes. Three PDGF ligand genes are expressed in the tumor endothelial cells, while PDGF-B-R is expressed in tumor pericytes (Bergers et al., 2003).

3.4.2 The K14-HPV16 Model

The first pattern of upregulation of angiogenesis-inducer genes is evident during epidermal squamous carcinogenesis in *K14-HPV16* transgenic mice. This tumor model expresses human papilloma virus type 16 (*HPV16*) oncogene under control of the keratin-14 (*K-14*) promoter. These mice express the *HPV16 E6* and *E7* oncogenes in the basal cells of their squamous epithelia (Arbeit et al., 1994) in the FVB/n strain background. They spontaneously develop epidermal squamous cell cancers (SCC) in a multistage fashion (Coussens et al., 1996). Their skin appears normal at birth, but becomes hyperplastic within the first month, and focal dysplasias develop between 3 and 6 months of age. These focal dysplasias are angiogenic and by 1 year have developed into invasive SCC in about half of the mice.

In normal murine epidermis, and in "normal" epidermis expressing the *HPV16* oncogenes, neither aFGF nor VEGF is transcribed at detectable levels. Both well and moderately differentiated SCC arise from pathways beginning as hyperplasia and progressing through varying degrees of dysplasia. A perceptible increase in dermal capillary density is first apparent in the first-month hyperplastic stage. There is a striking increase in both the number and distribution of dermal capillaries in the early and advanced dysplastic lesions; numerous vessels become closely apposed to the basement membrane separating dysplastic keratinocytes from the underlying stroma. The pattern is indicative of an angiogenic switch from vascular quiescence to the modest angiogenesis seen in the early, low-grade lesions, followed by a second, striking upregulation of angiogenesis in high-grade neoplasias as well as invasive cancer.

Progression is accompanied by the upregulation of proangiogenic factors, such as VEGF (Smith-Mc Cune et al., 1997) and FGF-2 (Arbeit et al., 1994), and the model has called attention to the involvement of proteases from inflammatory mast cells (Coussens et al., 1999), neutrophils, and macrophages in angiogenesis and tumor progression (Coussens et al., 2000). Arbeit et al. (1996) elaborated a derivative model of estrogen-induced carcinoma in female *K14-HPV16* mice. Cervical carcinoma developed in several stages in 80% of these mice after 6 months of estrogen treatment. Giraudo et al. (2004) used the *K14-HPV 16* transgenic tumor model to demonstrate MMP-9 in the tumor stroma concomitant with the angiogenic switch, expressed by infiltrating macrophages.

Neither the oncogene *SV40TAG* used in the creation of the *RIP-TAG* mice, nor oncogenes E6 and E7 used in the creation of *K14-HPV16* mice induced angiogenesis on their own. Additional changes during progression enable angiogenesis. All three oncogenes, in fact, bind and inactivate the tumor-suppressor gene products *p53* and *RB*.

3.4.3 The Papilloma Virus Type 1 Model

In this model, formation of dermal fibrosarcomas in *BPV1.69* transgenic mice occurred in three histologically distinct stages (normal, mild, and aggressive fibromatosis) (Sippola-Thiele et al., 1988) characterized by differential expression of *c-jun* and *jun-b* proto-oncogenes and their associated AP1 transcription factor activities (Bossy-Wtzel et al., 1992). Evaluation of microvascular density revealed a dramatic increase in capillaries in the aggressive fibromatosis stage (Kandel et al.,

1991). Moreover, conditioned medium from cells derived from normal dermis and mild fibromatosis did not display endothelial cell mitogenic activity, whereas those from the aggressive fibromatosis and fibrosarcoma did (Kandel et al., 1991). FGF-2 gene expression was at a similar level in all cell lines, irrespective of the presence or absence of mitogenic activity in conditioned medium. Thus, changes in FGF-2 gene expression did not explain the angiogenic switch.

Chapter 4

4.1 Tumor Endothelial Cell's Features

Endothelial cells of mature, quiescent vessels are characteristically low proliferative and their estimated turnover times are measured in years, whereas those of tumor vessels are markedly dependent on growth factors for survival. In contrast to tumor cells that are, by nature, genetically instable and heterogeneous, endothelial cells are normal diploid cells that do not acquire mutations, and therefore should not become resistant to therapy. In fact, although increasing evidence demonstrates the occurrence of genetic alterations, such as chromosomal translocations and aneuploidy in tumor-associated endothelial cells (Straubel et al., 2004; Hida et al., 2004; 2008), these cells are still considered to be more genetically stable than tumor cells, as they are not oncogenically transformed.

VEGF has been convincingly assigned a central role in the induction of host vessels into a growing tumor. When endothelial cells invade a newly formed tumor, they come into contact with tumor cells that produce VEGF, which may be responsible not only for vascular proliferation, but also for the altered permeability of the newly formed vessels.

Tumor endothelial cells proliferate 50–200 times faster than normal endothelial cells (Vermeulen et al., 1995). Their constant proliferation rate in some regions of the tumor vasculature reflects the angiogenesis that accompanies an increase in tumor volume, whereas in other regions they undergo apoptosis in parallel with tumor necrosis and vessel regression.

Tumor blood vessels are often abnormal being characterized by increased permeability, tortuosity, excessive random branching, and intratumoral variations in vascular lumen size (Table 4.1). Heterogeneous vascular morphology has been described in various tumor types, in tumors from the same origin growing in different host environment and in different stages of tumor progression (Table 4.2).

The tumor-associated endothelium is structurally defective. Discontinuities or gaps ($<2\ \mu$m in diameter) that allow hemorrhage and facilitate permeability are common features. Cell contacts are usually poorly differentiated and no complex contact structures exist. Defects in endothelial cell barrier function, due to abnormal cell–cell junctions and other changes, exaggerate leakiness. This correlates with histological grade and malignant potential (Daldrup et al., 1998) and can

D. Ribatti, *History of Research on Tumor Angiogenesis*,
DOI 10.1007/978-1-4020-9563-4_4, © Springer Science+Business Media B.V. 2009

Table 4.1 Differential features between normal and tumor vascular endothelium

Normal vessel	Tumor vessel
Normal organization	Abnormal organization
Normal pericyte	Pericyte is absent or detached
Normal basement membrane thick	Basement membrane is absent or too
Reliance on tubulin to maintain integrity is not required	Reliance on tubulin to maintain integrity is required
Vessel diameter has a normal distribution	Vessel is dilated
Vascular density is normal with a homogeneous distribution	Vascular density is abnormal with a heterogeneous distribution
Vascular permeability is normal	Vascular permeability is increased
pO$_2$ tension is normal	pO$_2$ tension is low
Drug penetration is uniform	Drug penetration is heterogeneous

be exploited in locating tumors by imaging contrast media and in the delivery of macromolecular therapeutics (Mc Donald and Choyke, 2003). Furthermore, it results in extravasation of plasma proteins and even erythrocytes and may facilitate the traffic of tumor cells into the bloodstream and the formation of metastases (Dvorak et al., 1998). Leakiness has been attributed to highly active angiogenesis and microvascular remodeling, but its structural basis and mechanism are unclear. Intercellular gaps, transendothelial holes, vesiculo-vacuolar organelles (VVO), and endothelial fenestrae are all present in the endothelium of tumor vessels (Dvorak et al., 1988).

Subpopulations of tumor cells with different angiogenic activities are present within one tumor. Isolation and subsequent subcutaneous or intradermal inoculation of these subpopulations into mice gave rise to tumors with different microvascular densities, apoptotic rates, and growth characteristics (Achilles et al., 2001; Yu et al., 2001).

Tumor endothelial cells also differ from those of normal vessels in other ways, including the profile and level of cell adhesion molecule they express. They preferentially overexpress the cell-surface molecules such as integrin $\alpha v\beta 3$ and $\alpha v\beta 5$, E-selectin, endoglin, endosialin, and VEGFR, all of which stimulate endothelium adhesion and migration (Magnussen et al., 2005). Tumors treated with antagonists of integrin $\alpha v\beta 3$ display increased apoptosis and vessel regression. One feature of the role of integrin $\alpha v\beta 3$ is that it mediates endothelial cell adhesion to extra-

Table 4.2 Specific tumor endothelial cells (EC) abnormalities

Brain tumor EC express lower levels of factor-VIII and proliferate more slowly than normal EC

Renal carcinoma tumor EC upregulate VEGF-D, Ang-1, and Akt and survive without serum

Melanoma and liposarcoma-derived EC have multiple centromes and are aneuploid

B-cell lymphoma-derived EC harbor lymphoma-specific genetic alterations

Breast tumor-derived EC are resistant to vincristine and doxorubicin

High-grade glioma-derived EC proliferate more rapidly than low-grade glioma EC

cellular matrix molecules. Vascular endothelial cadherin (VE-cadherin) is poorly expressed in tumor vessels. It results in their destabilization and may lead to abnormal remodeling.

Tumor endothelial cells present many abnormalities in gene expression. St Croix et al. (2000) found that tumor-activated endothelium overexpressed specific transcripts as a result of qualitative differences in gene profiling compared with the normal endothelial cells of the tissue of origin. A total of 79 transcripts were differentially expressed: 46 were elevated at least 10-fold and 33 were expressed at substantially lower levels in tumor-associated endothelial cells.

4.2 Tumor Basement Membrane

Vascular basement membrane marked by immunohistochemical staining for type IV collagen, laminin, fibronectin, or nidogen (entactin), is present on all normal blood vessels and on most blood vessels in tumors.

Vascular basement membrane and extracellular matrix are locally degraded to allow underlying endothelial cells to migrate into the perivascular space toward chemotactic angiogenic stimuli. Basement membrane is also a site of growth factor binding as a participant in angiogenesis (Kalluri, 2003).

The basement membrane that envelops endothelial cells and pericytes of tumor vessels may have extra layers that have no apparent association with the cells (Baluk et al., 2003). Spaces between the layers of type IV collagen, CD31, and alpha smooth actin reflect the loose association of endothelial cells and pericytes. Projections of basement membrane away from tumor vessels may accompany endothelial sprouts or pericyte processes (Baluk et al., 2003). Tumor basement membrane contains distinctive forms of fibronectin comprising the ED-B domain, and type IV collagen with exposed cryptic sites (Santamaria et al., 2003), and is a source of angiogenic and antiangiogenic molecules. Some of its structural proteins are broken down by enzymes to yield molecules with potent actions. Three examples are endostatin, which is a COOH-terminal fragment of collagen XVIII, tumstatin, and arresten, which are the noncollageneous-1 domain of the $\alpha 3$ and the $\alpha 1$ chains of type IV collagen, respectively (Colorado et al., 2000).

4.3 Tumor Pericytes

Pericytes around capillaries and SMCs around larger vessels, provide structural strength and participate in the maturation of the vasculature by controlling several endothelial cell functions (Table 4.3). Pericytes restrict the proliferation of endothelial cells in co-culture, where a single pericyte could contact and inhibit the growth of up to the endothelial cells (Ordlidge and D'Amore, 1987). This inhibition, requiring contacts between endothelial cells and pericytes, has been attributed to the activation of TGF-β (Antonelli-Orlidge et al., 1989). PDGF-B is involved

Table 4.3 Pericyte function in specific tissues

Brain
Blood–brain barrier
Immunological defense
Phagocytic activity

Liver
Vitamin A metabolism
Tissue repair

Kidney
Glomerular vascular function

Eye
Retinal vascular flow and function

in endothelial cell to pericyte signaling, stimulating pericyte migration and proliferation (Lindahl et al., 1997). A current hypothesis suggests that loosening of endothelial cell/pericyte contact may be necessary for the initiation of angiogenesis in mature adult vessels (Hanahan, 1997). In a hypoxic condition VEGF can also act as a pericyte mitogen (Yamagishi et al., 1999). Moreover, hypoxia promotes the in vitro growth of pericytes through the autocrine action of VEGF induced in this cell type.

Pericytes may differentiate into SMC. A study of mesenteric capillary growth in rats (Rhodin and Fuiita, 1989) suggests that fibroblasts transform into pericytes which, in turn, become SMC. Targeted disruption of the PDGF-B gene resulted in a defective development of the SMCs (Leeven et al., 1994).

Pericytes or SMCs may contribute to tumor neovascularization, as demonstrated in the microvasculature of glioblastoma multiforme, where many α-SMC actin-positive cells, reacting also with an antibody against activated pericytes, are detectable (Wesseling et al., 1995). Examination of TGF-β positive cells in Kaposi's sarcoma reveals the precursors in both SMCs and pericytes as well as in the spindle-shaped Kaposi's sarcoma cells (Williams et al., 1995). They are revealed by immunohistochemical staining of sections (Morikawa et al., 2002) and may cover 73–92% of endothelial sprouts in several murine tumor types. Breast and colon tumors recruit significantly more pericytes than gliomas or renal cell carcinoma (Eberhard et al., 2000).

Most tumor pericytes are loosely associated with endothelial cells, have abnormal shape, paradoxically extend cytoplasmic processes away from the vessel wall in the tumor parenchyma, and have extra layers of loosely fitting basement membrane (Morikawa et al., 2002).

Pericyte abnormalities are consistent with alterations in PDGF signaling pathways. Mice genetically deficient in PDGF-B or its receptors have blood vessels with loose pericyte attachment, irregular vessel caliber, luminal projections on endothelial cells, and hemorrhage (Betsholtz, 2004). Genetic abolition of PDGF-B receptor (PDGF-B-R) expressed by embryonic pericytes decreased their recruitment (Abramsson et al., 2003).

Similar abnormalities occur in many tumor vessels. PDGF-B expressed by tumor cells increases pericytes recruitment in several in vivo models (Guo et al., 1985). Ectopic expression of PDGF-B in tumor cells has been shown to facilitate an increase in pericyte density, but is unable to cause pericyte to attach more firmly to the blood vessels.

Pericytes were validated as a therapeutic target by using kinase inhibitors selective for the PDGF-B-R. Ablation of pericytes with a neutralizing antibody against PDGF-B-Rβ in pancreatic islet tumors of the *RIP1-TAG2* transgenic mice caused hyperdilation of tumor vessels and increased endothelial cell apoptosis (Song et al., 2005). Similarly, the receptor tyrosine kinase (RTK) inhibitor SU6668, which also affects PDGF-B-Rβ signaling, detached and diminished pericytes in *RIP1-TAG2* and xenotransplant tumor and thereby restricted the tumor growth (Reinmuth et al. 2001).

Clinical trials that are currently ongoing or in development aim to target endothelial cells and pericytes simultaneously and assess the potential benefits for antitumoral efficacy.

In Lewis lung carcinoma implanted in mice, inhibition of endothelial differentiation gene-1 (*EDG-1*) expression in endothelial cells strongly reduced the pericytes coverage (Chae et al., 2004). In a human glioma model developed in rat, Ang-1 led to the enhanced pericytes recruitment and increased the tumor growth, presumably by favoring angiogenesis (Machein et al., 2004). Alternatively, in a colon cancer model, overexpression of Ang-1 led to smaller tumors with fewer blood vessels and greater pericyte coverage, decreased vascular permeability, and reduced hepatic metastasis (Ahmad et al., 2001). Thus, depending on the model, stabilization of blood vessels by Ang-1 may either promote tumor angiogenesis or reduce tumor growth.

In a human neuroblastoma xenotransplanted model, pericytes coverage is halved in tumors grafted on MMP-9-deficient mice and the transplantation with MMP-9-expressing bone marrow cells restores the formation of mature vessels (Chantrain et al., 2004). In addition, overexpression of TIMP-3, a natural inhibitor of MMPs, results in decreased pericytes recruitment in neuroblastoma and melanoma models (Spurbeck et al., 2002). The observation that pericytes express MMPs in many human tumors in vivo (Nielsen et al., 1997) suggests that pericyte invasion requires the proteolytic degradation of extracellular matrix by proteases, including MMP.

Although inhibition of VEGF signaling pathways can lead to vessel regression, a few distinctive functional vessels remain that are slim and tightly covered with pericytes (Bergers et al., 2003). This observation suggests that endothelial cells can induce pericyte recruitment to protect themselves from death consequent to the lack of survival signals conveyed by VEGF. Accordingly, tumor vessels lacking adequate pericyte coverage are more vulnerable to VEGF inhibition (Bergers et al., 2003). Moreover, tumor pericytes express VEGF that supports endothelial cell survival (Darland et al., 2003).

Chapter 5

5.1 Tumor Lymphangiogenesis

The lymphatic system plays several roles in helath and disease (Table 5.1). Lymphatic vessels were first described in the beginning of the seventeenth century, when the Italian anatomist Gasparo Aselli identified lymphatic vessels as "milky veins" in the mesentery of a "well-fed" dog (Asellius, 1627). The developmental origin of lymphatic vessels remained unclear until Florence Sabin proposed in 1902 that endothelial cells bud off from the veins during early embryonic development and form primitive lymph sacs. This model was challenged in 1910 by Huntington and McClure who alternatively suggested that lymph sacs arise, independently of the veins from mesenchymal precursor cells (lymphangioblasts) with consecutive establishment of venous connections (Huntington and Mc Clure, 1910). Lymphangioblasts exist in birds and amphibians (Ny et al., 2005; Wilting et al., 2006). Although the existence of lymphangioblasts in mammals is unclear, it is possible that in addition to dedifferentiation of lymphatic endothelial cells from venous endothelial cells and subsequent sprouting lymphangiogenesis, differentiation of lymphatic endothelial cells from mesenchymal precursor cells may contribute to the formation of the lymphatic vasculature during embryonic development.

Table 5.1 Roles of the lymphatic system in health and disease

Health
Drains lymph fluid from the extracellular spaces
Absorbs lipids from the intestinal tract
Maintains the fluid homeostasis
Transports white blood and antigen-presenting cells to lymphoid organs
Disease
Congenital and acquired defects of the lymphatic system
Lymphedema
Lymphangitis
Lymphangioma
Lymphangiosarcoma
Metastasis

D. Ribatti, *History of Research on Tumor Angiogenesis*,
DOI 10.1007/978-1-4020-9563-4_5, © Springer Science+Business Media B.V. 2009

Detailed descriptive studies revealed the mechanisms of lymphangiogenesis in different animal and human tissues. Clark and Clark (1932) documented the extension of lymphatic capillaries by the outgrowth from existing lymph vessels in rabbit ear transparent chambers. Bellman and Odén (1959) documented via contrast microlymphangiography the time course and the extent of newly formed lymphatics in circumferential wounds of the rabbit ear, including lymphatic bridging through newly formed scar tissue (Odén, 1960). Subsequent studies documented the restoration of distinctive ultrastructural features in newly regenerated lymphatic vessels, including the characteristic overlapping junctional complexes and Weibel–Palade bodies (Magari and Asano, 1978; Magari, 1987).

Among the markers exhibiting specificity for lymphatic endothelium (Table 5.2), two membrane proteins seem to be particularly useful: LYVE1 is a transmembrane hyaluronic acid receptor-1 (Jackson et al., 2001) and podoplanin is a glomerular podocyte membrane protein (Breiteneder-Geleff et al., 1999). Furthermore, a homeobox protein, PROX-1, provides a nuclear marker that is expressed in lymphatics and not in blood vessels (Mouta-Carreira et al., 2001). PROX-1 knockout embryos completely lack lymph sacs and lymphatic vessels (Wigle and Oliver, 1999). Immunohistochemical analysis of VEGFR-3, LYVE-1, and podoplanin expression has shown the presence of lymphatic vessels in tissues that were previously thought to lack them, including various types of tumor, both experimental and human (Jackson et al., 2001).

The first growth factors and molecular marker specific for lymphatic vessels were discovered only 10 years ago. VEGF-C was described for the first time in 1996 by Joukov et al. and was characterized by the presence of unique amino- and carboxy-terminal extensions flacking the common VEGF-homology domain. VEGF-D was first isolated in 1996 from a different display screening of murine fibroblasts genes mice carrying a targeted inactivation of the *c-fos* gene (Orlandini et al., 1996). Transgenic mice overexpressing VEGF-C or VEGF-D in the skin show hyperplasia of cutaneous lymphatic vessels (Jeltsch et al., 1997; Veikkola et al., 2001), whereas VEGF-C null mouse embryos completely lack a lymphatic vasculature and die prenatally of fluid accumulation within the tissues (Karkkainen et al., 2004).

Henry Le Dran (1684–1770), a French surgeon, first proposed the theory that cancer begins in its earliest stages as a local disease, which initially spreads to the lymph nodes and subsequently enters the blood circulation. He also observed that the cure was much less likely when lymph nodes were involved.

Table 5.2 Markers exhibiting specificity for lymphatic endothelium within the vasculature

Desmoplakin
D6 (β-chemokine receptor)
5′-Nucleotidase
LYVE1
Podoplanin
PROX1
VEGFR-3

Cancer cells escape a tumor by two primary routes, blood and lymphatic vessels, to establish distant metastases. While lymph node metastases are not directly responsible for cancer-related death, cancer cells could spread from the lymph nodes to distant organs, where they can develop a secondary tumor and perturb critical functions of that organ. Although the metastatic dissemination of tumor cells to regional lymph nodes is a common feature of many human cancers, the exact mechanism whereby tumor cells enter the lymphatic system and whether tumors utilize existing lymphatic channels or tumor dissemination requires the de novo formation of lymphatics (lymphangiogenesis) is not clear (Clarijs et al., 2001; Pepper, 2001).

It is generally thought that the lymphatic invasion only occurs once the infiltration of tumor cells occurs upon pre-existing peritumoral lymphatic vessels. Historically, only few studies have addressed the issue of whether tumor lymphangiogenesis occurs at all, and only recently novel molecular markers for the lymphatics provided evidence for tumor lymphangiogenesis.

VEGF-C and VEGF-D have been identified as specific lymphangiogenic factors, which bind to and induce tyrosine phosphorylation of VEGFR-3 (Makinen et al., 2001). VEGF-C and VEGF-D also bind to NRP-2, a semaphorin receptor in the nervous system that is also expressed in lymphatic capillaries (Karkkainen et al., 2001). Consistent with these findings, NRP-2 deficient mice have lymphatic hypoplasia, which, however, regenerates during the postnatal period, whereas larger lymphatic vessels develop normally (Yuan et al., 2002).

Direct evidence for the role of VEGF-C and VEGF-D in lymphangiogenesis and metastasis was obtained by using mouse models. In a xenograft model of human breast cancer transplanted into immunodeficient mice, VEGF-C overexpression induced both peritumoral and intratumoral lymphangiogenesis, and increased metastasis to regional lymph nodes and to lungs (Skobe et al., 2001a). The degree of tumor metastases appears to correlate with intratumoral lymphatic vessel invasion into the tumors. VEGF-D-overexpressing epithelioid tumors induced the formation of intratumoral lymphatic vessels and promoted lymph node metastases (Stacker et al., 2001). In one study (Mandriota et al., 2001) transgenic (*RIP*-VEGF-C) mice that express VEGF-C specifically in pancreatic β cells and that, accordingly develop a lymphatic network around the β cells, were mated with a second transgenic (*RIP1-TAG2*) strain, which characteristically develop non-metastatic pancreatic β cell tumors. The double transgenic mice displayed VEGF-C-induced lymphangiogenesis around the β cell tumors, in addition to metastatic spread of tumor cells to pancreatic and regional lymph nodes. These data demonstrate that VEGF-C-mediated activation of the lymphatic system is sufficient to induce metastasis to lymph nodes in previously non-metastatic tumor cells.

VEGF-C overexpression led to lymphangiogenesis and growth of the draining lymphatic vessels, intralymphatic tumor growth, and lymph node metastasis in several tumor models (Saharinien et al., 2004), and clinicopathological studies have reported that expression of VEGF-C, VEGF-D, or VEGFR-3 can correlate with lymph node metastasis in human cancer, including malignant melanoma and lung, breast, colorectal, and gastric carcinomas (Table 5.3) (Stacker et al., 2002). Both

Table 5.3 Expression of VEGF-C/VEGF-D in tumors

VEGF-C
Breast cancer
Cervical cancer
Colorectal cancer
Endometrial carcinoma
Esophageal cancer
Gastric cancer
Head and neck squamous cell carcinoma
Lung adenocarcinoma
Neuroblastoma
Non-small cell lung cancer
Pancreatic cancer
Prostate cancer
Thyroid tumors
VEGF-D
Colorectal cancer
Lung adenocarcinoma
Head and neck squamous cell carcinoma

Table 5.4 Potential inhibitors of the lymphangiogenic signaling pathway

Soluble extracellular domain of VEGFR-3
Soluble extracellular domain of VEGFR-2
Neutralizing monoclonal antibodies to VEGF-C and VEGF-D
Neutralizing monoclonal antibodies to VEGFR-3
Peptidomimetics based on VEGF-C and/or VEGF-D
Inhibitors of tyrosine kinase activity of VEGFR-3

lymphangiogenesis and the formation of lymph node metastases can be inhibited by antagonists of VEGF-C and VEGF-D (Table 5.4) (Cao et al., 2005). In some models, lymphatics, but not lung metastases were blocked with the VEGF-C/D trap, whereas in others the treatment inhibited both types of metastases (He et al., 2002).

Nagy et al. (2002) have demonstrated that in addition to angiogenesis, VEGF-A also induces proliferation of lymphatic endothelium, resulting in the formation of greatly enlarged and poorly functioning lymphatic channels. At least, some of the effects of VEGF-A on lymphatic vessels may be secondary to the induction of vascular hyperpermeability and to the recruitment of the inflammatory cells that produce VEGF-C and VEGF-D (Cursiefen et al., 2004).

Most tumors have few or no functional lymphatic vessels. He et al. (2004) have shown, in fact, that their growth around tumors requires VEGFR-3 signaling. Few lymphatics are present in mice with heterozygous loss of function mutation of VEGF-C, and none are present when VEGFR-3 signaling is blocked. Blocking antibodies against VEGFR-3 decrease tumor angiogenesis and growth (Laakkonen et al., 2007).

Isaka et al. (2004) showed that lymphatics around tumors are functionally abnormal, and some may not drain in the correct direction. Skobe et al. (2001b) have shown that lymphatic capillaries activated by VEGF-C promote tumor cell invasion by increasing the tumor cell transendothelial migration. Proliferating intratumoral lymphatic vessels are present in certain human cancers, such as melanomas, head and neck carcinomas, and xenograft tumor models overexpressing lymphangiogenic factors (Daddras et al., 2003; Maula et al., 2003).

Although these findings provide no information on the mechanism of tumor cell dissemination, they raise the possibility that VEGF-C and VEGF-D may increase metastasis by increasing the number and size of lymphatic vessels, or alternatively by altering the functional properties of existing lymphatics. Nevertheless, the definitive role of VEGF-C and VEGF-D expression in promoting lymphangiogenesis in human tumors, and whether this leads to higher propensity for metastasis, remains to be established.

Understanding the mechanisms of lymphatic metastasis, including the identification of stromal and tumor determinants that are important for the spread of tumor cells through lymphatic vessels, represents an important challenge for tumor vascular biology researches.

Chapter 6

6.1 The Contribution of Inflammatory Cells to Tumor Angiogenesis

It was Rudolf Virchow in 1863, who critically recognized the presence of inflammatory cells infiltrating neoplastic tissues and first established a causative connection between the "lymphoreticular infiltrate" at sites of chronic inflammation and the development of cancer.

Epidemiological studies have shown that chronic inflammation predisposes individuals to various types of cancer. It is estimated that underlying infections and inflammatory responses are linked to 15–20% of all deaths from cancer worldwide (Balkwill and Mantovani, 2001). Accordingly, treatment with non-steroidal and anti-inflammatory agents decreases the incidence of, and the mortality that results from, several tumor types (Kochne and Dubois, 2004; Flossmann and Rothwell, 2007). Otherwise, marked chronic inflammatory response such as that in psoriasis is not associated with an increased risk of developing skin cancer (Nickoloff et al., 2005). The presence of inflammatory cells may be associated with better prognosis and inflammatory cells can destroy tumor cells (Mantovani et al., 1992). In 1893, William Coley noted that some patients with cancer who had severe postoperative infections at the tumor site underwent spontaneous and sustained tumor regression. He then developed Coley's mixed toxins, a filtrate from cultures of *Streptococcus pyogenes* and *Serratia marcescens* which was administered into the tumor or the surrounding tissues in patients with a range of advanced cancers and documented the cases of the long-term survival of individuals with malignancies.

In 1986, Dvorak noted that wound healing and tumor stroma formation shared many important properties (Dvorak, 1986). They both were rich of newly formed vessels and fibrin matrix. He likened tumors to "wounds that do not heal," because tumor cells secreted a vascular permeability factor, now referred to as VEGF, that could lead to persistent extravasation of fibrin and fibronectin and continuous generation of extracellular matrix. It is well established that tumor cells are able to secrete pro-angiogenic and pro-inflammatory factors as well as mediators for inflammatory cells. These later produce indeed cytokines such as VEGF, FGF-2, IL-8, PlGF, TGF-β, PDGF, Angs, and others (Table 6.1). These are exported

D. Ribatti, *History of Research on Tumor Angiogenesis*,
DOI 10.1007/978-1-4020-9563-4_6, © Springer Science+Business Media B.V. 2009

Table 6.1 Angiogenic factors released by inflammatory cells

Neutrophils	VEGF, IL-6, IL-8
Macrophages	VEGF, TGF-β, FGF-2, PDGF, substance P, prostaglandins, angiogenin
Mast cells	VEGF, FGF-2, TGF-β, TNF-α, IL-8, histamine, NGF, tryptase, heparin
Eosinophils	VEGF, FGF-2, TNF-α, TGF-β, PDGF, GM-CSF, NGF, IL-6, IL-8, eotaxin

from tumor cells or mobilized from the extracellular matrix. As a consequence, tumor cells are surrounded by an infiltrate of inflammatory cells, namely lymphocytes, neutrophils, macrophages, and mast cells. These cells communicate via a complex network of intercellular signaling pathways, mediated by surface adhesion molecules, cytokines, and their receptors. Results point to the importance of a cross-talk between several host cells for promoting angiogenic effects in tumor areas. Inflammatory cells cooperate and synergize with stromal cells as well as malignant cells in stimulating endothelial cell proliferation and blood vessel formation (Fig. 6.1). These synergies may represent important mechanisms for tumor development and metastasis by providing efficient vascular supply and easy pathway to escape. Indeed, the most aggressive human cancers, such as malignant melanoma, breast carcinoma, and colorectal adenocarcinoma, are associated with a dramatic host response composed of various inflammatory cells, especially macrophages and mast cells.

6.1.1 Neutrophils

Besides promoting inflammation-associated angiogenesis, there is increasing evidence that neutrophils also play a critical role in angiogenesis related to tumor progression (Di Carlo et al., 2003; Benelli et al., 2002). In fact, it is thought that tumor cells secrete factors that elicit a wound-repair response from neutrophils and tumor-associated macrophages and this response inadvertently stimulates tumor progression (Whalen, 1990). Neutrophils are recruited to sites of tumor growth by α-chemokines or CXC chemokines. Bellocq et al. (1998) showed a positive correlation between levels of IL-8 in patients with bronchoalveolar carcinoma and number of tumor-infiltrating neutrophils, suggesting that IL-8 secreted by these tumors recruits neutrophils to tumor sites. Furthermore, mice with subcutaneous ovarian tumors engineered to overexpress human IL-8 exhibit increased numbers of murine tumor neutrophils (Lee et al., 2000). In breast cancer, release by tumor-associated and tumor-infiltrating neutrophils of oncostatin M, a pleiotropic cytokine belonging to the IL-6 family, promotes tumor progression by enhancing angiogenesis and metastases (Queen et al., 2005). In addition, neutrophil-derived oncostatin M induces VEGF production from cancer cells and increases breast cancer cell detachment and invasive capacity (Queen et al., 2005). In a mouse model of endometriosis, early tissue infiltration with neutrophils promotes angiogenesis as well as recruitment and activation of macrophages which, in turn, amplify the angiogenic signal for the growth of endometriotic tissue (Lin et al., 2006). Expression of *HPV*

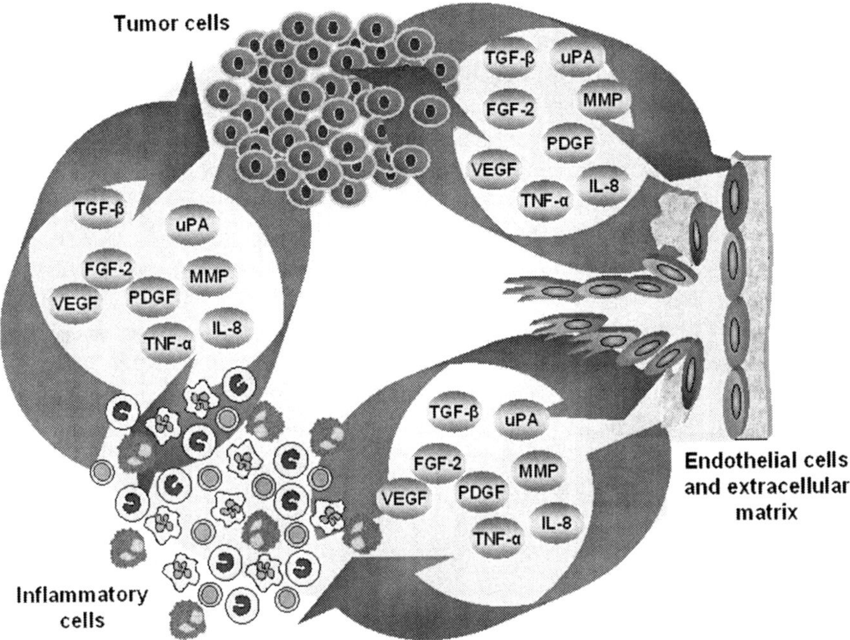

Fig. 6.1 Interplay between tumor cells, inflammatory cells, and extracellular matrix in inducing angiogenic response (Reproduced from Ribatti D, Vacca A, Overview of angiogenesis during tumor progression. In "Angiogenesis. An Integrative Approach from Science to Medicine", Figg WJ, Folkman J, eds., Springer Science, New York, 2008, pp. 161–8.)

16 early region genes in basal keratinocytes of transgenic mice elicits a multi-stage pathway to squamous carcinoma. Infiltration by neutrophils and mast cells, and activation of MMP-9 in these cells coincided with the angiogenic switch in premalignant lesions (Coussens and Web, 1996). In some tumor tissues, like melanoma, neutrophils are not a major constituent of the leukocyte infiltrate, but they might have a key role in triggering and sustaining the inflammatory cascade, providing chemotactic molecules for the recruitment of macrophages and other inflammatory and stromal cells. Neutrophils produce and release high levels of MMP-9. By contrast, neutrophils secrete little, if any, MMP-2, which plays an important role in the turnover of various extracellular matrix components (Muhs et al., 2003). However, neutrophils release a not yet identified soluble factor as well as a specific sulfatase and a heparanase that activate latent MMP-2 secreted by other cells and allow releasing of embedded growth factors from the extracellular matrix (Schwartz et al., 1998; Bartlett et al., 1995). Remodeled matrix facilitates the escape of tumor cells leaving the tumor mass to metastasize at distance, because it offers less resistance. In addition, proteolytic enzymes released by neutrophils can diminish cell–cell interactions and permit the dissociation of tumor cells from the original tumor site (Shamamian et al., 2001). The *RIP1-TAG2* transgenic mouse model has been used to further establish the role of neutrophils in tumor angiogenesis. The angiogenic islets

that formed within this early lesions contained increased number of neutrophils, and neutrophil depletion using anti-Gr1 significantly reduced the number of islets that underwent angiogenesis (Nozawa et al., 2006).

It is worth mentioning that neutrophils, besides releasing many pro-angiogenic molecules, also produce important antiangiogenic factors. Human neutrophils, for instance, synthesize and secrete from azurophilic granules small anti-microbial peptides known as α-defensins, which exert inhibition of endothelial cell proliferation, migration, and adhesion, impair capillary tube formation in vitro, and reduce angiogenesis in vivo (Chavakis et al., 2004). In addition, neutrophil-derived elastase can generate the antiangiogenic factor angiostatin (Scapini et al., 2002), a well-known inhibitor of IL-8-, macrophage inflammatory protein (MIP)-2-, and growth-related oncogen α (GRO-α)-induced angiogenesis in vivo (Benelli et al., 2002). Remarkably, *all-trans* retinoic acid, a promising molecule with potential antiangiogenic use in clinical treatment, has been shown to inhibit VEGF formation in cultured neutrophil-like HL-60 cells (Tee et al., 2006).

6.1.2 Monocytes and Macrophages

Cells belonging to the monocyte–macrophage lineage are a major component of the leukocyte infiltration in tumors and there is growing evidence that they are part of inflammatory circuits that promote tumor progression, favoring invasion and metastasis (Mantovani et al., 1992; Balkwill and Mantovani, 2001). Tumor-associated macrophages derive from circulating monocytes, which enter the neoplastic tissue through blood vessels. A number of tumor-derived chemoattractants ensures macrophage recruitment, including colony-stimulating factor-1 (CSF-1), the CC chemokines CCL2, CCL3, CCL4, CCL5, and CCL8, and VEGF secreted by both tumor and stromal elements (Mantovani et al., 2002). Besides killing tumor cells once activated by IFN-γ and IL-12, tumor-associated macrophages produce several pro-angiogenic cytokines as well as extracellular matrix-degrading enzymes that promote tumor progression and metastasis (Naldini and Carraro, 2005). These data are consistent with the formulated hypothesis of the "macrophage balance," which is to be connected with the dual opposite functions provided by macrophages in relation to neoplastic cell fate (Mantovani et al., 1992; Balkwill and Mantovani, 2001). According to the new experimental observations, indeed, tumor-associated macrophage density at the site of tumor development is critical for either destruction of the tumor mass or promotion of tumor progression. In a non-tumorigenic melanoma model, it has been shown that low level of macrophage infiltration supported tumor production while massive macrophage accumulation was associated with tumor destruction (Nesbit et al., 2001). The stimulating effect exerted by tumor-associated macrophages on the growth of the tumor mass is partly related to the angiogenic potential of these cells. The possibility that macrophages might be capable of modulating angiogenesis was first proposed by Sunderkotter and co-workers in 1991 (Sunderkotter et al., 1991). Today, an increasing body of evidence supports the view that tumor-associated macrophages are actually important elements in the complex process of regulating the formation of new vessels during the development and spread of malignant tumors. In the tumor microenvironment, macrophages are mainly represented by polarized type II (alternatively activated) or M2 elements, which would derive from tumor-associated macrophages upon local exposure to IL-4 and IL-10 (Mantovani et al., 2002). These cells have poor attitude to destroy tumor cells but are better adapted for promoting angiogenesis, repairing and remodeling wounded or damaged tissues, and suppressing adaptive immunity (Sica et al., 2006). Tumor-associated macrophages represent a rich source of potent pro-angiogenic cytokines and growth factors, such as VEGF, TNF-α, IL-8, and FGF-2. Macrophages express VEGFR-1 and VEGFR-3 and are thus attracted by angiogenic and lymphangiogenic signals. They also stimulate further lymphangiogenesis by secreting VEGF-C and VEGF-D (Skobe et al., 2001; Cursiefen et al., 2004; Schoppmann et al., 2002). In addition, these cells express a broad array of angiogenesis-modulating enzymes, including MMP-2, MMP-7, MMP-9, MMP-12, and cyclooxygenase-2 (COX-2) (Sunderkotter et al., 1991; Lewis et al., 1995; Klimp et al., 2001). The great number of pro-angiogenic factors secreted by macrophages may promote tumor spread and help explain the correlation between increased tumor-associated macrophage density and the augmented tumor vasculature recognized during experimental and human carcinogenesis. In humans, a significant relationship between the number of tumor-associated macrophages and the density of blood vessels has been established in tumors like breast carcinoma (Leek et al., 1995), melanoma (Makitie et al., 2001), glioma (Nishie et al., 1999), squamous cell carcinoma of the esophagus (Koide et al., 2004), bladder carcinoma (Hanada et al., 2000), and prostate carcinoma (Lissbrant et al., 2000). In the mouse cornea model, killing of COX-2 positive infiltrating macrophages with clodronate liposomes reduces IL-1β-induced angiogenesis and partially inhibits VEGF-induced angiogenesis (Nakao et al., 2005). In one model of subcutaneous melanoma, both angiogenesis and growth rate correlate with tumor infiltration by macrophages that express angiotensin I receptor and VEGF (Egami et al., 2003). In addition, Lewis lung carcinoma cells expressing IL-1β develop neovasculature with macrophage infiltration and enhance tumor growth in wild-type but not in monocyte chemoattractant protein-1 (MCP-1)-deficient mice, suggesting that macrophage involvement might be a prerequisite for IL-1β-induced neovascularization and tumor progression (Nakao et al., 2005). Remarkably, in a murine model of mammary carcinoma, deficiency of macrophage-colony stimulating factor (M-CSF), a potent inductor of macrophage recruitment in tumor tissues, does not affect early stages of tumor development but reduces progression to invasive carcinoma and metastasis (Lin et al., 2001). This result highlights the possible role of tumor-associated macrophages in contributing to the angiogenic switch that accompanies transition into malignancy. In Polyoma Middle T antigen-induced mouse mammary tumors, indeed, focal accumulation of macrophages in premalignant lesions precedes the angiogenic switch and the progression into invasive tumors. Depletion of tumor-associated macrophages reduces to about 50% of tumor vascular density, leading to areas of necrosis by loss of blood supply within the tumor mass. Reintroduction of macrophages into these knockout mice led to a significant increase in vascular density and enhanced tumor progression (Lin et al., 2006).

Macrophages have been shown to accumulate particularly in such necrotic and hypoxic areas in different neoplasia, like human endometrial, breast, prostate, and ovarian carcinomas (Ohno et al., 2004; Leek et al., 1999). It is otherwise known, indeed, that upregulation of the pro-angiogenic program in tumor-associated macrophages, followed by increased release of VEGF, FGF-2, TNF-α urokinase, and MMPs, is stimulated by hypoxia and acidosis (Bingle et al., 2002). Moreover, activated macrophages synthesize and release inducible nitric oxide synthase, which increases blood flow and promotes angiogenesis (Jenkins et al., 1995).

Hypoxic macrophages respond to hypoxia by upregulating hypoxia-inducible transcription factors, mainly HIF1 and HIF2 (Burke et al., 2002; Talks et al., 2000), the activation of which leads to increased transcription of many genes that regulate cell proliferation, metabolism, and angiogenesis (Lewis and Murdoch, 2005).

The angiogenic factors secreted by macrophages stimulate migration of other accessory cells that potentiate angiogenesis, in particular mast cells (Gruber et al., 1995). Macrophages have been considered "as obligate partners for tumor cell migration, invasion, and metastasis" (Condeelis and Pollard, 2006). It has recently been recognized that osteopontin, a phosphorylated acidic glycoprotein (Denhardt et al., 2001) deeply affects the pro-angiogenic potential of human monocytes. Osteopontin presents either as an immobilized extracellular matrix component or a soluble molecule. Reports suggest that osteopontin may affect angiogenesis by acting directly on endothelial cells and/or indirectly via mononuclear phagocyte engagement, enhancing the expression of TNF-α and IL-1β in mononuclear cells (Leali et al., 2003; Naldini et al., 2006).

It should also be mentioned that monocytes and macrophages are primary producers of IL-12. This multifunctional cytokine can cause tumor regression and reduce metastasis in animal models, due to the promotion of anti-tumor immunity and also to the significant inhibition of angiogenesis (Colombo and Trincheri, 2002). The anti-angiogenic activity is mediated by IFN-γ production, which in turn induces the chemokine IFN-γ-inducible protein (IP)-10 (Angiolillo et al., 1995; Romagnani et al., 2001). There is in vitro evidence that IL-12 inhibits VEGF produced by breast cancer cells and regulates stromal cell interactions, leading to decreased MMP-9 and increased tissue inhibitor of metalloproteinase (TIMP)-1 production (Dias et al., 1998). Thus, macrophages have the potential to either stimulate angiogenesis and promote tumor progression or inhibit angiogenesis and cause tumor regression.

De Palma et al. (2005) discovered a subpopulation of circulating blood monocytes called Tie-2-expressing monocytes (TEM) for their expression of the angiopoietin receptor-2. TEM are found in human tumors including those of the kidney, colon, pancreas, and lung as well as in soft tissue sarcomas (Venneri et al., 2007) and have also been identified in spontaneous and orthotopic tumors in mice (De Palma et al., 2005). Moreover, Venneri et al. (2007) showed that human U87 gliomas grown subcutaneously in nude mice were significantly more vascularized in mice injected with human Tie2$^+$ monocytes than in mice injected with monocytes lacking Tie2 expression.

6.1.3 Lymphocytes

Tumor-infiltrating lymphocytes may contribute to cancer growth and spread, and to the immunosuppression associated with malignancies. The predominant T cell population homing in the tumor microenvironment has a "memory" phenotype, while NK cells are rare (Negus et al., 1997). T cells infiltrating various types of human tumors express a Th2 polarization (type II phenotype) and predominance of Th2 cells may be a general strategy to subvert immune responses against tumor cells. CD8$^+$ cytotoxic T cells predominate in Kaposi's sarcoma while CD4$^+$ Th cells prevail in cervical carcinoma (Balkwill and Mantovani, 2001). By producing IL-4, IL-13, and IL-10 but not IFN-γ, tumor-infiltrating T cells might reinforce the skewing of monocyte differentiation in tumors toward an M2 phenotype (Mantovani et al., 2002). Polarized Th2 cells are indeed ineffective against tumor cells and viruses. Tumor-associated lymphocytes also display defective signaling via the T-cell receptor (Mizoguchi et al., 1992).

By contrast, lymphocytes may cooperate to the generation of an antiangiogenic microenvironment that is essential for causing regression of the tumor mass. For instance, Th cells and cytotoxic T cells are needed to mediate the antiangiogenic effect of IL-12 (Strasly et al., 2001). In addition, experimental work suggests that NK cells are required as mediators of angiogenesis inhibition by IL-12, and that NK cell cytotoxicity of endothelial cells is a potential mechanism by which IL-12 can suppress neovascularization (Yao et al., 1999). IL-12 receptors indeed are present primarily on NK cells and T cells (Trichieri, 1993). IL-12-activated lymphocytes influence inhibition of tumor growth and function as an anti-vascular agent, by releasing higher level of IFN-γ and down-modulating VEGF (Cavallo et al., 2001).

6.1.4 Eosinophils

Increased numbers of eosinophils have been described in many human tumors including nasopharyngeal (Looi, 1987) and oral squamous cell carcinoma (Dorta et al., 2002), gastrointestinal tumors (Nielsen et al., 1999), and Hodgkin lymphoma (Teruya-Feldstein et al., 1999). Eosinophils are thought to be recruited to tumors, in part by the selective chemoattractant CCL11 (eotaxin) which binds to CCR3 on these cells (Daugherty et al., 1996). Recent evidence suggests that eosinophils recruited to tumor sites can influence angiogenesis. Freshly isolated human blood eosinophils or supernatants from cultured eosinophils induce endothelial cell proliferation in vitro and angiogenesis in the rat aortic ring assay and CAM assay, suggesting that eosinophils can directly influence angiogenesis (Puxeddu et al., 2005a). Eosinophils contain preformed VEGF in their secretory granules, which is rapidly secreted upon activation with IL-5 (Horiuchi and Weller, 1997). Eosinophils also release other proangiogenic factors like FGF-2, IL-6, IL-8, GM-CSF, PDGF, and TGF-β (Puxeddu et al., 2005b).

6.1.5 Mast Cells

Experimental evidence indicates mast cells as key host cells in the tumor infiltrate, with important consequence on tumor cell fate. On the one hand, mast cells express detrimental effects on tumor growth by producing molecules killing tumor cells and inducing an inflammatory reaction. On the other hand, mast cells favor tumor progression by promoting expansion of its vascular supply, degradation of the tumor extracellular matrix, and immunosuppression (Theoharides and Conti, 2004). Mast cells accumulate at sites of tumor growth in response to numerous chemoattractants, like small tumor-derived peptides (Kessler et al., 1976; Poole and Zetter, 1983), RANTES and MCP-1 (Conti et al., 1997).

Mast cells play a role in tumor growth and angiogenesis. Mast cells synthesize and release a vast array of pro-inflammatory and pro-angiogenic molecules that favor new vessel formation either directly or via local recruitment of activated in-flammatory cells. Mast cell-deficient W/W^v mice exhibit indeed a decreased rate of tumor angiogenesis (Starkey et al., 1988). Furthermore, bone-marrow transplants to repair this mast cell deficiency restored the full angiogenic response (Starkey et al., 1988). Coussens et al. (1999) showed that premalignant angiogenesis is decreased in *HPV-16*-derived cutaneous squamous cell carcinoma in mast cell-deficient trans-genic mice compared with their wild-type counterparts. Molecules like heparin could facilitate tumor vascularization not only by a direct pro-angiogenic effect but also through its anti-clotting effect (Theoharides and Conti, 2004). In addition, mast cell-derived MMPs can degrade the interstitial tumor stroma and hence release matrix-bound angiogenic factors. An increased number of mast cells have indeed been reported in angiogenesis associated with vascular neoplasms, like hemangioma and hemangioblastoma (Glowacki and Mulliken, 1982), as well as a number of solid and hematopoietic tumors. At present, it is not known if tumor-associated mast cells are involved in "sprouting" angiogenesis or in "non-sprouting" angiogenesis, or in both. In general, mast cell density correlates with angiogenesis and poor tumor out-come. Association between mast cells and new vessel formation has been reported in breast cancer (Harveit, 1981; Bowrey et al., 2000), colorectal cancer (Lachter et al., 1995), and uterine cervix cancer (Graham and Graham, 1966). Tryptase-positive mast cells increase in number and vascularization increases in a linear fashion from dysplasia to invasive cancer of the uterine cervix (Benitez-Bribiesca et al., 2001). An association of VEGF and mast cells with angiogenesis has been demonstrated in laryngeal carcinoma (Sawatsubashi et al., 2000) and in small lung carcinoma, where most intratumoral mast cells express VEGF (Imada et al., 2000; Takanami et al., 2000; Tomita et al., 2000). Mast cell accumulation has also been noted around melanomas, especially invasive melanoma (Reed et al., 1996; Dvorak et al., 1980). Mast cell accumulation was correlated with increased neovascularization, mast cell expression of VEGF (Toth et al., 2000) and FGF-2 (Ribatti et al., 2003b), tumor aggressiveness, and poor prognosis. Indeed, a prognostic significance has been attributed to mast cell and microvascular density not only in melanoma (Ribatti et al., 2003c) but also in squamous cell cancer of the esophagus (Elpek et al., 2001). Recently, angiogenesis has been shown to correlate with tryptase-positive mast cell count in human endometrial cancer. Both parameters were found to increase in agreement with tumor progression (Ribatti et al., 2005a).

Mast cell density, new vessel rate, and clinical prognosis have also been found to correlate in hematological tumors. In benign lymphadenopathies and B cell non-Hodgkin's lymphomas, angiogenesis correlates with total and tryptase-positive mast cell counts, and both increase in step with the increase with Working Formulation malignancy grades (Ribatti et al., 1998; 2000). In non-Hodgkin's lymphomas, a correlation has been found between vessel count and the number of mast cells and VEGF-expressing cells (Fukushima et al., 2001). In the bone marrow of patients with inactive and active multiple myeloma as well as those with monoclonal gammopathies of undetermined significance, angiogenesis highly correlates with mast cell counts (Ribatti et al., 1999a). A similar pattern of correlation between bone marrow microvessels' count, total and tryptase-positive mast cell density, and tumor progression has been found in patients with myelodysplastic syndrome (Ribatti et al., 2002) and B cell chronic lymphocytic leukemia (Ribatti et al., 2003a). In the early stages of B cell chronic lymphocytic leukemia, the density of tryptase-positive mast cells in the bone marrow has been shown to predict the outcome of the disease (Molica et al., 2003).

As mast cells have the potency to express either favorable or detrimental effects on tumor cell growth, the hypothesis has been proposed that such dual role may depend on the way mast cells release their bioactive molecules from secretory granules. Frank exocytosis would export secretory cytokines mainly involved in promoting tumor cell apoptosis while piecemeal degranulation, a particulate and possibly selective way of mast cell secretion, would allow for release of mediators and growth factors principally responsible for angiogenesis, immunosuppression, and extracellular matrix disruption (Theoharides and Conti, 2004). Interestingly, lymph node and bone marrow mast cells in B cell non-Hodgkin's lymphomas and multiple myeloma show ultrastructural features of slow and particulate secretion as it occurs in piecemeal degranulation (Ribatti et al., 1999; 2000; Crivellato et al., 2002; 2003).

Chapter 7

7.1 The Role of Endothelial Progenitor Cells in Tumor Angiogenesis. The First Isolation of Putative Endothelial Progenitor Cells

In 1997, Asahara et al. reported the isolation of putative endothelial progenitor cells from human peripheral blood, on the basis of cell-surface expression of CD34 and other endothelial markers. They reported that hematopoietic cells differentiate into endothelial cells in vitro and in vivo. Moreover, the data suggested that hemangioblast, a common hematopoietic and endothelial cell precursor, might be present in the blood (Asahara et al., 1997). These cells were reported to differentiate in vitro into endothelial cells and seemed to be incorporated at sites of active angiogenesis in various animal models of ischemia.

7.2 Characterization of Endothelial Progenitor Cells

Asahara et al.(1997) used a polyclonal antibody to the intracellular domain of VEGFR-2 to show that CD34$^+$, and VEGFR-2$^+$ circulating endothelial progenitor cells form colonies that take up acetylated LDL. When CD34$^+$, VEGFR-2$^+$, CD34$^-$, or VEGFR-2$^-$ cells were injected into mice, rats, and rabbits undergoing neovascularization due to hind limb ischemia, CD34$^+$ and VEGFR-2$^+$ cells, but rarely CD34$^-$ or VEGFR-2$^-$ cells, incorporate into the vasculature in a manner consistent with their being endothelial cells.

A novel hematopoietic stem cell marker, AC133, a five-transmembrane glycoprotein whose function is unknown, is also expressed on endothelial precursor cell subsets, but not on mature endothelial cells (Yin et al., 1997). Its expression is rapidly downregulated as hematopoietic progenitors and endothelial progenitor cells differentiate (Yin et al., 1997; Miraglia et al., 1997).

Gehlin et al. (2000) demonstrated that AC133$^+$ cells from granulocyte colony stimulating factor (G-CSF)-mobilized peripheral blood differentiate into endothelial cells when cultured in the presence of VEGF and stem cell growth factor. Phenotypic analysis revealed that most of these cells display endothelial features, including the

D. Ribatti, *History of Research on Tumor Angiogenesis,*
DOI 10.1007/978-1-4020-9563-4_7, © Springer Science+Business Media B.V. 2009

expression of VEGFR-2, Tie-2, and vWF. All these data indicate that AC133 is currently the best selective marker for identifying endothelial progenitor cells and that circulating $CD34^+$, $VEGFR-2^+$, and $AC133^+$ cells constitute a phenotypically and functionally distinct population of circulating endothelial progenitor cells that may play a role in postnatal vasculogenesis.

Peichev et al. (2000) showed that a small subset of $CD34^+$ cells from different hematopoietic sources express both AC133 and VEGFR-2. Incubation of this subset with VEGF, FGF-2, and collagen results in their proliferation and differentiation into $AC133^-$, $VEGFR-2^+$ mature endothelial progenitor cells. Maturation and in vitro differentiation of these cells abolish AC133 expression, suggesting that endothelial progenitor cells with angioblast potential may be marked selectively with AC133.

On the basis of this characterization, endothelial progenitor cells can be roughly subdivided into three inter-related classes. One class of endothelial progenitor cells is represented by cells that are likely to be hemangioblasts. In mice this includes $Sca-1^+$ $c-kit^+$ lin^- cells and $Sca-1^+$ $c-kit^+$ $CD34^-$ cells from the bone marrow (Jackson et al., 2001; Grant et al., 2002) and in humans, $CD113^+$, $CD34^+$, and $VEGFR-2^+$ cells from the bone marrow and blood (Asahara et al., 1997; Peichev et al., 2000; Schatteman et al., 2000; Salven et al., 2003; Nowak et al., 2004).

Romagnani et al. (2005) have shown that cells that are positive for CD34 marker alone (and AC133 or CD14 markers) also express Nanog and Oct-4, which are embryonic stem cell markers. These cell populations are capable of proliferation in response to stem cell growth factors and differentiate into endothelial cells, adipocytes, osteoblasts, and neural cells.

Two different endothelial progenitor cell subpopulations have been described, denoted as early and late endothelial progenitor cells, with distinct cell growth patterns and ability to secrete angiogenic factors (Gulati et al., 2003; Hur et al., 2004). Early endothelial progenitor cells are spindle-shaped cells, which have a peak growth in culture at 2–3 weeks and which die after approximately 4 to weeks in vitro and secrete an array of angiogenic, antiangiogenic, and neuroregulatory cytokines (Hur et al., 2004). Late endothelial progenitor cells are cobblestone shaped and usually appear after 2–3 weeks of culture, show exponential growth at 4 weeks, can be maintained for up to 12 weeks in culture (Gulati et al., 2003; Hur et al., 2004) and are developed exclusively from the $CD14^-$ subpopulations (Urbich et al., 2003).

7.3 Triggers and Inhibitors of Endothelial Progenitor Cells to Sites of Active Neovascularization

When endothelial progenitor cells were exposed to angiogenic factors, they formed highly proliferative endothelial colonies, whereas circulating endothelial cells could only generate endothelial monolayers that had limited proliferation capacity because they are mature, terminally differentiated cells (Solovey et al., 1997).

Endothelial progenitor cells' recruitment to sites of neoangiogenesis is triggered by the increased availability of angiogenic growth factors or chemokines, such as VEGF, angiopoietin, and stromal cell-derived factor (SDF)-1α (Hattori et al., 2001; Iwaguro et al., 2002; Yamaguchi et al., 2003). The latter binds to the chemokine receptor CXCR-4, which is highly expressed on endothelial progenitor cells (Mohle et al., 1998).

Release of endothelial progenitor cells from bone marrow into circulation can be induced by granulocyte macrophage colony stimulating factor (GM-CSF) or VEGF and is critically dependent on the activity of endothelial nitric oxide synthase (eNOS) expressed by stromal cells in bone marrow (Aicher et al., 2003). Once avoided at the sites of neovascularization, endothelial progenitor cells may recruit additional endothelial progenitor cells by releasing growth factors, such as VEGF, hepatocyte growth factor (HGF), G-CSF, and GM-CSF (Rehman et al., 2003).

Erythropoietin and estrogen also positively influence endothelial progenitor cell numbers (Heeschen et al., 2003; Strehlow et al., 2003). Vasa et al. (2001) demonstrated the positive effects of statins on endothelial progenitor cells such as increasing the number of circulating endothelial progenitor cells, reducing senescence, enhancing proliferation rate, and differentiation from CD34$^+$ cells.

Angiogenic factors activate MMPs, specifically MMP-9 (Vu and Werb, 2000), which lead to the release of soluble KIT ligand (sKITL) (Engsig et al., 2000) which, in turn, promotes the proliferation and motility within the bone marrow microenvironment, thereby laying the framework for endothelial progenitor cells mobilization to the peripheral circulation.

Certain inhibitors of tumor neovascularization may act by inhibiting mobilization and homing of endothelial progenitor cells to the developing vascular network of tumors. Consistent with this notion, the in vitro proliferation and colony-forming ability of human endothelial progenitor cells are markedly decreased in the presence of angiostatin, a proteolytic cleavage product of plasminogen with antiangiogenic properties (Ito et al., 1999).

7.4 The Transplantation Models to Study Endothelial Progenitor Cells

Shi et al. (1998) used a transplantation model in which marrow cells from donors and recipients were distinguishable to determine whether endothelial cells lining a vascular prosthesis are derived from the bone marrow. After 12 weeks, the graft was retrieved, and cells with endothelial morphology were identified; only donor alleles were detected in DNA from positively stained cells on the graft. These results suggested that a subset of CD34$^+$ cells located in the bone marrow mobilize to the peripheral circulation and colonize the endothelial flow surface of vascular prostheses. In those studies, grafts were prepared such that their flow surface was accessible to cells from blood, but shielded from external pannus and perigraft endothelial ingrowth.

Asahara et al. (1999 a) employed a bone marrow transplant model in mice, sublethally irradiated to demonstrate the incorporation of bone marrow endothelial progenitor cells into foci of neovascularization. Four weeks after transplants, when the bone marrow had been reconstituted, a variety of surgical experiments to provoke neovascularization were performed: (i) Cutaneous wounds examined 4 and 7 days after skin removal by punch biopsy disclosed a high frequency of incorporated endothelial progenitor cells. (ii) One week after the onset of hind limb ischemia, endothelial progenitor cells were incorporated in capillaries among skeletal myocytes. (iii) After permanent ligation of a coronary artery, myocardial infarction sites demonstrated incorporation of endothelial progenitor cells into the foci of neovascularization at the border of the infarct. All these findings clearly indicate that postnatal neovascularization does not rely on angiogenesis alone and that bone marrow endothelial progenitor cells contribute to postnatal vasculogenesis.

Lin et al. (2000) found a distinction between vessel wall and bone marrow-derived endothelial cells in blood samples from subjects who had received gender-mismatched bone marrow transplants 5–20 months earlier. They showed that 95% of circulating endothelial cells had recipient genotype and 5% had donor genotype. After 9 days of culture, endothelial cells derived predominantly from the recipient vessel wall, expanded only 6-fold, compared with 98-fold after 27 days by endothelial cells, mostly originated from donor bone marrow cells. These data suggest that most of the circulating endothelial cells in fresh blood originate from vessel walls and have limited growth capability, and that outgrowth of endothelial cells is mostly derived from transplantable marrow-derived cells.

Kalka et al. (2000) demonstrated that transplantation of endothelial progenitor cells to athymic mice with hind limb ischemia markedly improves the blood flow recovery and capillary density in the ischemic limb and significantly reduces the rate of limb loss.

7.5 The Identification of Bone Marrow-Derived Multipotent Progenitor Cells

Reyes et al. (2002) have identified a single cell in human and rodent postnatal marrow that they term bone marrow-derived multipotent progenitor cells. Multipotent progenitor cells were selected by depleting adult bone marrow of hematopoietic cells expressing CD45 and glycophorin-A, followed by long-term culture on fibronectin with epidermal growth factor (EGF) and PDGF under low serum conditions. A cell population expressing AC133 and low levels of VEGFR-1 as well as VEGFR-2 and the embryonic stem cell marker Oct-4 emerged. Its culture with VEGF induced differentiation into CD34$^+$, vascular endothelial-cadherin, VEGFR-2$^+$ cells, a phenotype consistent with angioblasts. Subsequently these cells express vWF and markers of mature endothelium, such as CD31, CD36, and CD62-P. They form vascular tubes when plated on Matrigel and upregulate angiogenic receptors and VEGF in response to hypoxia. In immunocompetent mice, intravenously injected endothelial cells contribute to neovascularization of transplanted tumors and participate in wound healing. Despite extensive efforts, hematopoietic differentiation of human multipotent progenitor cells has not been observed, indicating a significant difference from the pathway by which both hematopoietic cells and endothelial cells are generated from hemangioblasts.

Differentiation of human multipotent progenitor cells toward the endothelial lineage was only induced by seeding them at a high density in serum-free medium with the addition of VEGF, whereas culture in medium with fetal calf serum directed the differentiation into osteoblasts, chondroblasts, and adipocytes (Reyes et al., 2001).

7.6 Monocyte/Macrophage as a Source of Endothelial Progenitor Cells

Peripheral blood mononuclear cells from adult humans can be enriched in endothelial progenitor cells by the addition of VEGF, FGF-2, insulin-like growth factor, and EGF to the culture medium for 7–10 days. After local injection in vivo, these cells contribute to the formation of new vessels in the ischemic limb (Kalka et al., 2000).

Rehman et al. (2003) demonstrated that endothelial progenitor cells isolated from the monocyte/macrophage fraction of peripheral blood did not proliferate, but secreted angiogenic growth factors. Monocytic endothelial progenitor cells may enhance the angiogenic process via a release of inflammatory mediators that stimulate granulation tissue formation.

Urbich et al. (2003) demonstrated that $CD14^+$ cells purified from peripheral blood mononuclear cells and cultured in endothelial growth media form adherent cells that have characteristics of endothelial cells, such as expression not only of vWF and VEGFR-2, but also of CD45.

7.7 Contribution of Endothelial Progenitor Cells to Tumor Angiogenesis

The importance of endothelial progenitor cells is dependent on the percent of vessels within the vasculature that incorporate endothelial progenitor cells. Preclinical models of tumor growth are characterized by rapidly growing tumors in which the rate of incorporation of endothelial progenitor cells is relatively high, with 90% of the endothelial cells comprising the tumor vasculature in certain tumors (Rafii et al., 2002). In humans, the percent of endothelial progenitor cells incorporated is lower (12%), but sufficient and necessary for the conversion of avascular micrometastases to progressive metastatic tumors (Gao et al., 2008).

Tumor angiogenesis is supported by the mobilization and functional incorporation of endothelial progenitor cells. The recruitment of endothelial progenitor cells to tumor angiogenesis represents a multistep process, including (i) active arrest and homing of the circulating cells within the angiogenic microvasculature; (ii) transendothelial extravasation into the interstitial space; (iii) extravascular formation of cellular clusters; (iv) creation of vascular sprouts and cellular networks; (v) incorporation into a functional microvasculature.

The fact that endothelial progenitor cells can contribute to tumor angiogenesis indicates that endothelial progenitor cells, although are primarily programmed to form blood vessels during embryonic vascular development, retain this ability within an angiogenic environment in the adult. Endothelial progenitor cells have been detected at increased frequency in the circulation of cancer patients and lymphoma-bearing mice, and tumor volume and tumor production of VEGF were found to be correlated with endothelial progenitor cells mobilization (Mancuso et al., 2001; Monestiroli et al., 2001).

Vajkoczy et al. (2003) investigated the mechanisms of homing and incorporation of endothelial progenitor cells during new blood vessel formation in a tumor model using mouse embryonic endothelial precursor cells as a model system and they showed that endothelial progenitor cells retain their ability to contribute to tumor angiogenesis in the adult. Circulating endothelial progenitor cells are specifically arrested in "hot spots" within the tumor microvasculature, extravasate in the interstitium, form multicellular clusters, and incorporate into functional vascular networks.

Bone marrow-derived cells, including endothelial progenitor cells, restored impaired VEGF-driven angiogenesis in mice lacking PlGF (Carmeliet et al., 2001). PlGF null mice have impaired tumor angiogenesis due to decreased vessel density. When these mice received bone marrow transplantation from wild-type mice, they were able to support tumor angiogenesis, indicating that PlGF is delivered to tumor vasculature by bone marrow-derived cells.

The percentage of endothelial progenitor cells incorporation is generally low and depends on the nature of the tumor, supporting the concept that most tumor neovascularization seems to occur via angiogenesis (Table 7.1). However, in some model systems, tumors are reliant on endothelial progenitor cells mobilization (Lyden et al., 2001).

High levels of VEGF produced by tumors may result in the mobilization of bone marrow-derived stem cells in the peripheral circulation and enhance their recruitment into the tumor vasculature (Asahara et al., 1999b; Hattori et al., 2001).

Hypoxia can also mobilize endothelial progenitor cells from the bone marrow in the same way, as hematopoietic cytokines, such as GM-CSF (Takahashi et al., 1999). Malignant tumor growth results in neoplastic tissue hypoxia, and may mobilize bone marrow-derived endothelial cells in a paracrine fashion and thus contribute to the sprouting of new tumor vessels.

Table 7.1 Human tumors in which endothelial progenitor cells have been detected

Breast cancer
Lymphoma
Multiple myeloma
Acute myeloid leukemia
Myelodysplastic syndrome

When endothelial progenitor cells are engrafted into immunocompromised mice, they incorporate into the vasculature of xenotransplanted tumors. To distinguish between the contribution of bone marrow-derived endothelial progenitor cells and that of neighboring vessels many studies have taken advantage of transplantation techniques in which sex-mismatched or genetically marked donor bone marrow is infused into irradiated recipients. Id 1/3 double-mutant mouse embryos have vascular malformations in the forebrain, leading to fatal hemorrhage. Adult mice with reduced Id gene dosage cannot support tumor-induced angiogenesis (Lyden et al., 2001). Lyden et al. (2001) demonstrated that transplantation and engrafment of β-galactosidase-positive wild-type bone marrow or VEGF-mobilized stem cells into lethally irradiated Id-mutant mice is sufficient to reconstitute tumor angiogenesis. In contrast to wild-type mice, Id-mutants fail to support the growth of tumors because of impaired angiogenesis. Tumor analysis demonstrates uptake of bone marrow-derived VEGFR-2-positive endothelial progenitor cells into vessels surrounded by VEGFR-1-positive myeloid cells. Defective angiogenesis in Id-mutant mice is associated with impaired VEGF-induced mobilization and proliferation of the bone marrow precursor cells. Inhibition of both VEGFR-1 and VEGFR-2 signaling is needed to block tumor angiogenesis and induce necrosis. The mechanism for underlying angiogenesis defect is believed to be due, in part, to impaired recruitment of endothelial progenitor cells to the tumor. Both B6RV2 lymphoma and Lewis lung carcinoma cells, which fail to grow tumors in Id-deficient mice, are able to form fully vascularized tumors in Id-mutant mice that have received wild-type bone marrow transplants.

In another experimental approach human multipotential progenitor cells, which are highly primitive cells that have the capacity to differentiate into different cell types, were induced to differentiate into endothelial cells. Reyes et al. (2002) found that in vitro-generated multipotential progenitor cells respond to angiogenic stimuli by migrating to tumor sites and contributing to tumor vascularization. Multipotential progenitor cells were injected into immunocompromised mice that carried mouse Lewis lung carcinoma. After 5 days, 30% of the newly formed tumor-associated vessels were derived from human multipotential progenitor cells-derived endothelial cells. Reyes et al. (2002) also found that in vivo angiogenic stimuli in a tumor microenvironment are sufficient to recruit multipotential progenitor cells to the tumor bed and induce their differentiation into endothelial cells that contribute to the tumor vasculature. In fact, multipotential progenitor cells contributed to the neovasculature of spontaneously formed lymphomas that commonly develop in ageing immunocompromised mice.

Asahara et al. (1999a) subcutaneously injected murine colon cancer cells into mice that underwent bone marrow transplantation from transgenic mice constitutively expressing the β-galactosidase gene regulated by an endothelial cell-specific promoter. Three weeks after tumor implantation, histological examination of the tumors revealed multiple Lac-Z-positive cells both within the tumor stroma and incorporated within the endothelial layer of tumor blood vessels.

In a bone marrow transplantation model in which donor bone marrow was transduced with a lentiviral vector encoding green fluorescent protein (GFP) driven by the endothelial-specific Tie 2 promoter, it was estimated that only 0.05% of blood vessels in tumor xenografts were derived from bone marrow endothelial precursor cells (Garcia-Barros et al., 2003). These results, which conflict with multiple prior reports on contribution to tumor vasculogenesis, may be due to subtle differences in methodology. The percent contribution of endothelial precursor cells to tumor vasculature appears to depend heavily on the experimental model, and thus it remains imprecisely defined.

Shaked et al. (2006) demonstrated a strong correlation between tumor growth and endothelial progenitor cell numbers in mice using various tumor models and were able to effectively define the optimal antiangiogenic treatment (anti-VEGFR-2) based on endothelial progenitor cells monitoring.

It is important to point out that the role of bone marrow-derived cells in tumor angiogenesis remains controversial. Purhonens et al. (2008), in contrast with other studies, reported that bone marrow-derived VEGFR-2-positive cells or other endothelial progenitor cells do not contribute to the angiogenic tumor vasculature.

Chapter 8

8.1 Tumor Microvascular Density as a Prognostic Indicator

As early in 1972, Brem et al. proposed a microscopic angiogenesis grading system to assess the angiogenic status of the tumor vasculature. Based on the analysis of the vascular density, the number of endothelial cell nuclei, and endothelial cytology, an angiogenic score was determined and used to establish an angiogenic rank order of different human brain tumors.

In 1988, Srivastava et al. studied the vascularity of 20 intermediate thickness skin melanomas (0.76–4.0 mm level of invasion). Vessels were highlighted with *Ulex europaeus-1* agglutinin conjugated with peroxidase, and the stained histological sections were analyzed with a semi-automatic image analysis system. The ten cases that developed metastases had a vascular area at the tumor base that was more than twice that seen in the ten cases without metastases. Age, sex, Breslow's tumor thickness, and Clark's level of invasion were similar in the two groups.

In 1991, Weidner et al. developed a new method to perform microvascular density counting studies within tumors. The first step in Weidner's approach is the identification by light microscopy of the area of highest neovessel density, the so-called hot spot, by scanning the whole tumoral section at low power, then, individual microvessels are counted at a higher power ($\times$ 200 field) in an adequate area (e.g., 0.74 mm^2 per field using $\times$ 20 objective lens and $\times$ 10 ocular). The technique for identifying neovascular hot spots is very similar to that of finding mitotic hot spots for assessing mitotic figure content within breast tumors and is subjected to the same kind of inter- and intra-observer variability. Sclerotic, hypocellular areas within tumors, and immediately adjacent to benign breast tissues were not considered in microvascular density determinations. Any stained endothelial cell or clusters separate from adjacent vessels are counted as a single microvessel, even in the absence of vessel lumen. Vessels with muscular walls were not counted. Vessel lumen and red cells were not used to define a microvessel. Each single count is expressed as the highest number of microvessels identified at the hot spot. By using this approach, Weidner et al. (1991) showed that intratumoral microvascular density mammary tumors with poor prognosis and metastasis is twice as high in patients with mammary tumors

D. Ribatti, *History of Research on Tumor Angiogenesis*,
DOI 10.1007/978-1-4020-9563-4_8, © Springer Science+Business Media B.V. 2009

with good prognosis and without metastasis, and confirmed this correlation also in prostate carcinoma (Weidner et al., 1993). Other studies performed on different patients' databases by different investigators at different medical centers have observed the same association of increasing intratumoral vascularity with various measures of tumor aggressiveness, such as higher stage at presentation, greater incidence of metastases, and/or decreased patient survival.

Some authors used the Chalkley count which consists of applying a 25-point Chalkley eyepiece graticule (Chalkley, 1943) on several hot spots. Briefly, three or four areas of tumor are chosen. The graticule is applied to each hot spot using a specific magnification ($\times$ 250) with a corresponding defined Chalkley grid area ($0.196 \, mm^2$). The graticule is oriented to allow the maximum number of points to hit on or within the areas of stained microvessel profiles. This technique, suggested as a standard in an international consensus (Vermeulen et al., 1996), is considered to be a simple and acceptable procedure for daily clinical use (Fox, 1997; Hansen et al., 1998). Hansen et al. (2000) studied a population-based group consisting of 836 patients with operated primary, unilateral invasive breast carcinomas and demonstrated that there were significant correlations between high Chalkley counts and axillary lymph node metastasis, large tumor size, high histological malignancy grade, and histological type.

8.2 Use of Panendothelial Cell Markers

Interest in grading tumor angiogenesis was stimulated with the advent of non-specific endothelial markers, such as endogenous alkaline phosphatase and lectins, but only since the later part of 1980s, as more specific endothelial markers have become available, has the quantification been performed.

The use of different panendothelial cell markers may account for some of the variation in the estimation of microvascular density. When applied properly, anti-factor VIII-related antigen (RA) remains the most specific endothelial marker, providing very good contrast between microvessels and other tissue components (Fig. 8.1).

Although apparently more sensitive and superior on paraffin sections, CD31 (Fig. 8.2) strongly cross-reacts with plasma cells (De Young et al., 1993). This complication can markedly obscure the microvessels in those tumors with a prominent plasma cellular inflammatory background. Another disadvantage of CD31 staining is the frequent antigen loss due to fixatives containing acetic acid. Using anti-CD31 antibodies, regions with a prominent inflammatory infiltrate might be erroneously taken for a vascular hot spot at low magnification. These CD31 cell infiltrates sometimes obscure microvessels, especially single cell sprouts.

CD34 is an acceptable alternative and the most reproducible endothelial cell highlighter in many laboratories, but CD34 will highlight perivascular stromal cells and has been noted to stain a wide variety of stromal neoplasms (van de Rijn and Rouse, 1994; Traweek et al., 1991).

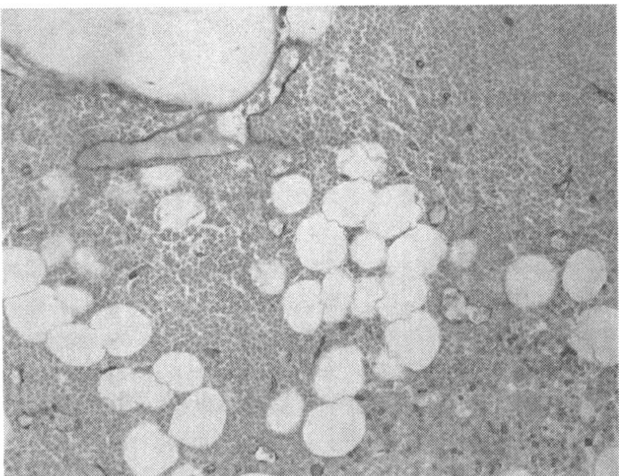

Fig. 8.1 Immunohistochemical reactivity for factor VIII-RA in B cell chronic lymphocytic leukemia bone marrow specimen

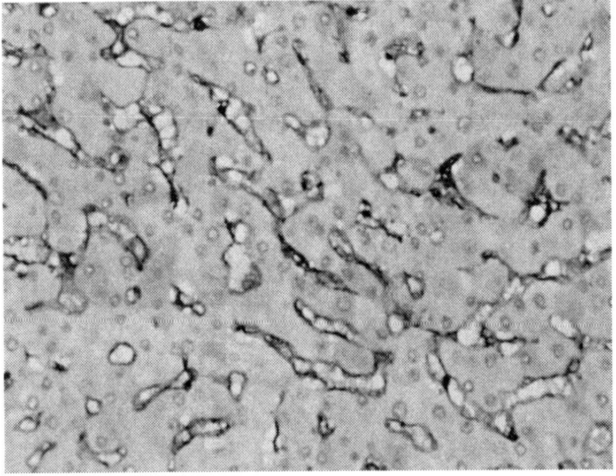

Fig. 8.2 Immunohistochemical reactivity for CD31 in human hepatocellular carcinoma specimen

It is noteworthy to emphasize that none of the above markers are able to discriminate between quiescent versus activated/proliferating endothelium. Two antibodies, namely E-9 (Wang et al., 1993; 1994) and CD105 (endoglin) (Kumar et al., 1999) seem to be specific for activated/proliferating endothelial cells.

CD105, a proliferation-associated and hypoxia-inducible protein, is preferentially expressed in the activated endothelial cells participating in neoangiogenesis, especially tumors, and is undetectable or weakly expressed in vessels of normal tissues (Kumar et al., 1999; Marioni et al., 2005, 2006; Minhajat et al., 2006; Sandlund et al., 2006).

Tie-2/Tek, an endothelium-specific receptor tyrosine kinase (Dales et al., 2004a) and VEGF receptors (Dales et al., 2004b) also identify stromal vessels.

8.3 Topography

Topography is important in differentiating tumor vessels into those supplying the invading tumor edge (i.e., the zone of tumor/normal tissue interaction) and those serving the inner tumor area.

The peripheral tumor areas are composed of typical capillaries with endothelial cells derived from pre-existing vessels. In contrast, the central areas of the tumor are made up of tube-like endothelial structures and pseudo-vascular channels lined by tumor cells, not endothelial cells. These tumor-lined spaces, commonly known as vasculogenic mimicry, are probably generated directly by the tumor cells rather than through the expression of angiogenic factors (non-angiogenesis-dependent pathway). The two vascular channels are inter-communicating (Ribatti et al., 2003d; Yue and Chen, 2005; Zhang et al., 2007).

Although areas of hot spots are not infrequently seen within the inner tumor area, they usually predominate at the edge of tumor. The vascular density was high in the tumor areas adjacent to normal tissues, but decreased gradually toward the inner tumor areas, although to a variable degree between different types of tumors.

As was pointed out by Thomlinson and Gray (1955) one can think of the supported tumor cells as forming a viable cuff around a vessel, with cuff size being roughly indicative of the metabolic burden of the cancer cells. Cuff size tends to vary inversely with tumor metabolic demand. Tumors that have high rates of oxygen or nutrient consumption, such as glioblastomas, have small cuffs only two or three cells wide and have a high vascular density. In contrast, tumors of low metabolic demand, such as chondrosarcomas, have a relatively large cuff size, with many cell layers supported and a relatively low vascular density.

8.4 Certain Human Tumor Types Can Exhibit Lower Microvascular Density than the Corresponding Normal Tissues

The measurement of microvascular density is not sufficient to reveal the functional or angiogenic status of tumor neovasculature. Microvascular density for human lung, mammary, renal cell, and colon carcinomas is lower than those of their normal tissue counterparts (Eberhard et al., 2000). In lung carcinoma, for example, microvascular density was found to be only 29% that of normal lung tissue. Microvascular density for glioblastomas was found to be 78% that of normal brain tissue and pituitary adenomas are less vascular than the normal pituitary gland (Turner et al., 2000). The apparent paradox can be partially explained by the lower oxygen consumption rate of tumor cells (Steinberg et al., 1997). All tumor vessels are not equal in their ability to provide oxygen and nutrients to the tumor cells they support. Tumor vessels can themselves be hypoxic and carry little oxygen, or they can have oscillating rather than directed blood flows and thus be ineffective at transporting oxygen and nutrients (Hickley and Simon, 2006; Brahimi-Horn and Pouyssegur,

2006; Gruber and Simon, 2006). Inhibiting hypoxic vasculature would shrink a tumor mass less than would inhibiting oxygen-rich vessels (Boyle and Travers, 2006).

In addition, tumor cells are known to tolerate oxygen deprivation and to be resistant to apoptosis under hypoxic conditions (Graeber et al., 1996). Both the lowered oxygen consumption of tumor cells and their tolerance of hypoxic conditions promote increased intercapillary distance in tumors relative to their normal counterparts.

8.5 Prognostic Value of Microvascular Density

Microvascular density would be a good indicator of therapeutic efficacy, but it has not been as useful for efficacy as it has been for prognosis. Since the early studies, hundreds of reports have examined the prognostic value of microvascular density in several forms of cancer. Most of these studies report positive correlation between microvascular density and tumor recurrence. Several studies on microvascular density and prognosis gave positive results in patients with solid tumors, such as head and neck, lung, gastric, colorectal, liver, pancreatic, renal, bladder, ovarian, endometrial and breast cancers, and neuroblastoma (Cernea et al., 2004; Chandrachud et al., 1997; Zhao et al., 2005; Lindmark et al., 1996; Lackner et al., 2004; Ribatti et al., 2006b; Couvelard et al., 2005; Yao et al., 2007; Korkolopoulou et al., 2001; Kusamura et al., 2003; Cantu De Leon et al., 2003; Shimizu et al., 2000; Fernandez-Aguilar et al., 2006; Ribatti and Ponzoni, 2005). In gliomas, microvascular density appears to correlate with the outcome in high-grade, but not with low-grade tumors, and does not correlate with tumor cellularity in the infiltrating portions of the tumors (Sharma et al., 2006). More recently, a positive correlation between microvascular density and tumor recurrence has been established also in hematological tumors (Korkolopoulou et al., 2001; 2005; Ridell and Norrby, 2001; Alexandrakis et al., 2004; Lundberg et al., 2006; Vacca and Ribatti, 2006). Nevertheless, despite the initial confirmatory publications, numerous reports appeared in the literature that failed to show a positive association between increasing tumor vascularity and reduced patient outcome, and caution as to the clinical utility of tumor angiogenesis is being urged (Bossi et al., 1995; Busam et al., 1995; Marrogi et al., 2000; Hillen et al., 2006). However, many of these negative studies may result from significant differences in methodologies.

8.6 The Relationship Between Microvascular Density and Intercapillary Distance

Intercapillary distance is determined at the local level by the net balance between angiogenic factors that stimulate and those that inhibit vessel growth, as well as by non-angiogenic factors, such as the oxygen and nutrient consumption rates of tumor cells (Yoshii and Sugiyama, 1988; Mikhail et al., 2004). In turn, microvascular density is determined by intercapillary distance, which in a tumor is dictated by the thickness of the perivascular cuff of tumor cells. Tumors that have high rates of oxygen and nutrient consumption have small cuffs only two to three cells wide and have a high vascular density (Gordan and Simon, 2007; Mabjeesh and Amir, 2007). On the contrary, tumors with low rates of oxygen consumption have relatively large cuff sizes and a relatively low vascular density (Gordan and Simon, 2007; Mabjeesh and Amir, 2007). This is an important parameter as it is the goal of an antiangiogenic tumor therapy to reduce the intercapillary distance to a such a degree that it becomes rate-limiting step for the growth of the tumor.

Chapter 9

9.1 Inhibitors of Angiogenesis

The existence of specific angiogenesis inhibitors was first postulated by Folkman in 1971, in an editorial published in the *"New England Journal of Medicine."* This paper was the first to employ the term "antiangiogenesis" to describe a potential therapeutic approach. No angiogenesis inhibitors existed before 1980 and few scientists thought at that time that such molecules would ever be found.

The attempt to discover angiogenesis inhibitors became possible after the development of bioassays for angiogenesis during the 1970s. Before, there were a few in vivo bioassays that had provided the majority of the earlier data in the field. These included the long-term culture of vascular endothelial cells, the development of the chick embryo CAM, and rabbit/murine cornea assays. In 1973, Gimbrone in Folkman's laboratory (Gimbrone et al., 1974) and Eric Jaffe's laboratory at Cornell (Jaffe et al., 1973) were independently the first to successfully grow and passage vascular endothelial cells from human umbilical veins in vitro. The first long-term passage of cloned capillary endothelial cells was reported in 1979 (Folkman et al., 1979). A further advance was achieved with the finding that the capillary endothelial cells could be induced to form three-dimensional networks in vitro that had same properties of capillary networks in vivo (Folkman et al., 1979).

Among the in vivo assays, the most widely used are the cornea and the CAM assays. Gimbrone in the Folkman's laboratory implanted tumor samples of approximately $0.5\,mm^3$ into the stromal layers of the rabbit cornea at a distance of up to 2 mm from the limbal edge. New capillary blood vessels grew from the limbus, invaded the stroma of the avascular corneas, and reached the edge of the tumor over a period of approximately 8–10 days (Gimbrone et al., 1974). One advantage of this assay was that it clearly showed the movement of new blood vessels through a tissue that had no background vessels to obscure the view. Langer and Folkman (1976) developed the use of polymeric systems that could deliver proteins continuously over an extended time period and were able to mix potential angiogenic factors into the polymeric release system and apply the mixture to the corneal pocket. The corneal assay has been pivotal for the development of new drugs that inhibit angiogenesis.

D. Ribatti, *History of Research on Tumor Angiogenesis,*
DOI 10.1007/978-1-4020-9563-4_9, © Springer Science+Business Media B.V. 2009

In its original from, the CAM assay was performed on 7–9-day-old chicken embryos by making a window in the egg shell, and then placing tissue or organ graft directly on the membrane (Ribatti et al., 2009). A variety of total embryo culture methods have been developed, beginning with a simple eversion into a Petri dish of the entire egg content at 3 days of incubation (Auerbach et al., 1974).

Beginning in the 1980s, the biopharmaceutical industry began exploiting the field of antiangiogenesis for creating new therapeutic compounds for modulating new blood vessel growth in angiogenesis-dependent diseases. From 1980 to 2005, 11 angiogenesis inhibitors were identified or discovered in the Folkman's laboratory (Table 9.1). The majority of these were found in the blood or tissues. Some of them were previously unknown molecules, such as angiostatin and endostatin, while for others, angiogenesis inhibition was a new function, such as interferon α and platelet factor 4. Other laboratories joined this research effort.

There are two general classes of angiogenesis inhibitors, direct and indirect (Table 9.2). A direct angiogenesis inhibitor blocks vascular endothelial cells from proliferating, migrating, or increasing their survival in response to proangiogenic proteins. Among direct angiogenesis inhibitors, there are angiostatin and endostatin. Both molecules were peptides that had been cleaved from larger naturally occurring, biologically active proteins: angiostatin from plasminogen and endostatin from collagen XVIII. This suggested that proteolytic enzymes produced by the tumor cells were crucial for the generation of angiogenesis inhibitors. Otherwise, indirect angiogenesis inhibitors decrease or block expression of a tumor cell product, neutralize the tumor product itself, or block its receptor in endothelial cells. Most indirect angiogenesis inhibitors are designed to target growth factor signaling pathways and can block the activity of one, two, or a broad spectrum of proangiogenic proteins and/or their receptors.

Clinical trials involving targeted antiangiogenic agents (Table 9.3) have demonstrated clinical efficacy and good tolerability, leading to the approval of the first targeted antiangiogenic drug, a humanized anti-VEGF monoclonal antibody, Avastin (bevacizumab). At present, 10 drugs with antiangiogenic activity have received Food

Table 9.1 Molecules with antiangiogenic activity discovered in the Folkman's laboratory between 1980 and 2005

Year	Molecule(s)
1980	Interferon alpha
1981	Platelet factor 4, protamine
1985	Angiostatic steroids
1990	TNP 470
1994	Angiostatin
1994	Thalidomide
1994	2-Methoxyestradiol
1997	Endostatin
1998	Cleaved antithrombin III
2002	3-Amino thalidomide
2005	Caplostatin

Table 9.2 Direct and indirect angiogenesis inhibitors

Direct
Angiostatin
Bevacizumab (Avastin)
Arresten
Canstatin
Combrestatin
Endostatin
Thrombospondin
Tumstatin
2-Methoxyestradiol
Vitaxin
Indirect
Targeting EGF-receptor tyrosine kinase
Targeting VEGF receptor
Targeting PDGF receptor
Targeting ERBB-2
Targeting interferon alpha receptor*

* Interferon alpha can be considered both a direct angiogenesis inhibitor, because it inhibits endothelial-cell migration, and an indirect angiogenesis inhibitor because it inhibits synthesis of FGF-2 by tumor cells.

Table 9.3 Angiogenesis inhibitors in clinical trials

Bevacizumab
Sunitinib
Sorafenib
GW786034B (pazopanib)
AZD2171
PKT787/ZK222584 (vatalanib)
VEGF-Trap
ZD6474
AMG706
AG013736

and Drug Administration approval in the USA and in more than 30 other countries by their regulatory agencies (Table 9.4).

9.1.1 Interferons

The antiproliferative activity of interferons (IFN) against human tumors was first demonstrated in the 1960s with partially purified IFN-α by Strander at the Karolinska Institute (Strander, 1986). A mixture of IFN inhibited the migration of capillary endothelial cells in vitro (Brouty-Boye and Zetter, 1980) and lymphocyte-induced angiogenesis in vivo (Sidky and Borden, 1987), as well as tumor angiogenesis

Table 9.4 Drugs with antiangiogenic activity which have received Food and Drug Administration approval for treatment of cancer

Date approval	Drug and disease
May 2003	Velcade (bortezomib)/multiple myeloma
December 2003	Thalidomide/multiple myeloma
February 2004	Avastin (bevacizumab)/colorectal cancer
February 2004	Erbitux/colorectal cancer
November 2004	Tarceva (erlotinib)/lung cancer
September 2005	Endostatin/lung cancer
December 2005	Nexavar (sorafenib)/kidney cancer
December 2005	Revlimid (lenalidomide)/myelodysplastic syndrome
June 2006	Revlimid/multiple myeloma
October 2006	Avastin/lung cancer
March 2007	Avastin/metastatic breast
May 2007	Torisel (CCI-779)/kidney cancer

(Dvorak and Gresser, 1989). The first angiogenesis inhibitor IFN-α administered at low doses was reported in 1980 (Brouty-Boye and Zetter, 1980). By 1995, Fidler showed that IFN-α shuts off overproduction of FGF-2 in human tumor cells (Singh et al., 1995). IFN-α has been reported to suppress the production of FGF-2 by human cancer cells (Slaton et al., 1999).

Since 1988, IFN-α has been used successfully to cause complete and durable regression of life-threatening pulmonary hemangiomatosis, hemangiomas of the brain, airway, and liver in infants, recurrent high-grade giant cell tumors refractory to conventional therapy, and angioblastomas (Ezekowitz et al., 1992; Kaban et al., 1999; 2002; Ginns et al., 2003). These tumors all express high levels of FGF-2 as their major angiogenic mediator. Low dose of daily IFN-α therapy for 1–3 years is sufficient to return abnormally high levels of FGF-2 in the urine of these patients to normal.

9.1.2 Cartilage

Certain tissues in the body have small number of vessels and appear to be relatively avascular. Folkman speculated that cartilage was avascular owing to the presence of naturally occurring angiogenesis inhibitors in this tissue.

In 1975, Brem and Folkman demonstrated that tumor-induced vessels were inhibited by a diffusible factor from neonatal rat cartilage (Brem and Folkman, 1975). Cartilage has been studied as a potential source of an angiogenesis inhibitor because of its avascularity. In fact, cartilage is a relatively tumor-resistant tissue and the tumor associated with cartilage, chondrosarcoma, is the last vascularized of all solid tumors. In 1980, Langer et al. partially purified extracts of cartilage, which inhibited tumor-induced neovascularization when delivered regionally (via controlled release polymer) and systemically (via infusion) (Langer et al., 1980). Ten years later Langer et al. purified an angiogenesis inhibitor from bovine scapular cartilage, obtained amino terminal sequence, and demonstrated that antiangiogenic activity could be attributed in part to the inhibition of metalloprotease activity (Moses et al., 1990).

9.1.3 Protamine and Platelet Factor 4

In 1982, Taylor and Folkman characterized protamine, a sperm-derived cationic protein, capable of inhibiting neovascularization in the chick embryo CAM assay and tumor growth and metastases when administered systemically, although its efficacy was limited to its toxicity at high doses (Taylor and Folkman, 1982).

Platelet factor 4 was first tested for antiangiogenic activity because its method of binding and neutralizing heparin is similar to that of protamine (Taylor and Folkman, 1982). Recombinant human platelet factor 4 (rHuPF4) has been produced (Maione et al., 1990) and it specifically inhibited endothelial proliferation and migration in vitro (Sharpe et al., 1990). The inhibitory activities are associated with the carboxy-terminal region of the molecule. The growth of human colon carcinoma in athymic mice, as well as the growth of murine melanoma, was markedly inhibited by intralesional injections, whereas tumor cells were completely insensitive to rHuPF4 in vitro at levels that inhibited normal endothelial cell proliferation. Systemic administration of rHuPF4 has so far been ineffective against tumor growth, perhaps because of the rapid inactivation or clearance of the peptide.

9.1.4 Vascular Disrupting Agents

In 1982, Denekamp hypothesized that the local disruption of the tumor vasculature would result in the death of many thousands of tumor cells, and that only a few endothelial cells within the vessels need to be killed to completely occlude the vessels (Denekamp, 1982). This strategy relies on the ability of the vascular disrupting agents (VDA) to distinguish the endothelial cells of tumor capillaries from normal ones based on their different phenotype, increased proliferative potential and permeability, and dependence on tubulin cytoskeleton (Table 9.5). Unlike angiogenesis inhibitors, vascular targeting agents should require only intermittent administration to synergize with conventional treatment rather than chronic administration. VDA work best in the poorly perfused hypoxic central tumor areas, leaving a viable rim of well-perfused cancer tissue at the periphery, which rapidly regrows (Tozer et al., 2005).

At present, there are three distinct classes of small molecule VDA undergoing clinical development: (i) tubulin depolymerizing agents; (ii) synthetic flavonoids; (iii) N-cadherin inhibitors. The largest group of low molecular weight VDA are the tubulin-binding combretastatin, namely CA4 and CA1, which are structurally related to the classical tubulin-binding agent, colchicine. Currently, three combretastatin derivatives are in clinical development, the phosphate prodrug of CA4

Table 9.5 Differential characteristics of antiangiogenic and vascular disrupting agents

Antiangiogenic agents
Inhibit tumor growth
Cytostatic
Optimal biologic dose < maximum tolerated dose
Long treatment period
Drug resistance is a concern
Specificity of target is less important
Targets must be functionally required for angiogenesis

Vascular disrupting agents
Cause rapid collapse of tumor blood flow and tumor necrosis
Cytotoxic
Optimal biologic dose = maximum tolerated dose
Rapid treatment period
May circumvent the development of drug resistance
Specificity of target is very important
Target molecules need only be correlated with angiogenesis

(OXi2021), Oxi4503, and AVE8062. Two other non-combretastatin-based tubulin depolymerizing agents, ZD6126 and MN-029, developed as VDA have entered into clinical trials. Combination of VDA and chemo and/or radiation therapy, which targets cancer cells at the tumor periphery, has produced promising responses in preclinical models.

9.1.5 Angiostatic Steroids

Folkman had begun to use the CAM of the chick embryo to detect angiogenic activity in fractions being purified from tumor extracts. The addition of heparin increased the speed of development of the angiogenic reaction so that it could be read 1–2 days later (Taylor and Folkman, 1982). But one problem with this assay is that occasionally eggshell dust falls on the CAM and causes background inflammation. Folkman guessed that adding cortisone to the CAM might eliminate the irritation from the shell dust but not abolish the tumor-angiogenic reaction. As expected, cortisone alone prevented shell dust inflammation without interfering with angiogenesis induced by tumor extracts. The surprise was that when heparin and cortisone were added together tumor angiogenesis was inhibited (Folkman et al., 1983). Furthermore, when this combination of heparin and steroid was suspended in a methylcellulose disk and implanted on the young (6-day) CAM, growing capillaries regressed leaving in their place, 48 h later, an avascular zone. The antiangiogenic effect was specific for growing capillaries. Mature non-growing capillaries in older membranes were unaffected. Non-anticoagulant heparin had the same effect. A hexasaccharide fragment with a molecular weight of approximately 16,000 was found to be the most potent inhibitor of angiogenesis (in the presence of a corticosteroid). The combination of the heparin hexasaccharide fragment and cortisone also inhibited tumor-induced angiogenesis in the rabbit cornea.

The regression of a growing vessel exposed to heparin–steroid combinations begins with endothelial cell rounding and is followed by cessation of endothelial proliferation, desquamation of endothelial cells, and retraction of the capillary sprout (Ingber et al., 1986). These events occur as 24–48 h and are accompanied by dissolution of the basement membrane of the new capillary vessels.

9.1.6 Fumagillin

Fumagillin was found by Ingber in the Folkman's laboratory to inhibit endothelial cell proliferation without causing endothelial cell apoptosis, when a tissue culture plate of endothelial cells became contaminated with a fungus *Aspergillus fumigatus fresenius* (Ingber et al., 1990). A conditioned medium from fungal cultures contained an inhibitor of endothelial cell proliferation and angiogenesis, which, upon purification, was found to be fumagillin, a polyene macrolide. When capillary endothelial cells were stimulated by FGF-2, half-maximal inhibition was observed with fumagillin at 100 pg/ml. This antiproliferative effect appeared to be relatively specific for endothelial cells because inhibition of non-endothelial cells, including tumor cells, was observed at up to 1,000-fold higher concentrations. Scientists at Takeda Chemical Industries (Osaka, Japan) made a synthetic analog of fumagillin, called TNP-470, which inhibits endothelial proliferation in vitro at a concentration of 3 logs lower than the concentration necessary to inhibit fibroblasts and tumor cells.

TNP-470 showed significant inhibition of tumors in clinical trials, including durable complete regression (Milkowski et al., 1999). The clinical utility of TNP 470, however, was limited by neurotoxicity. This side effect was overcome when Ronit Satchi-Fainaro in the Folkman's lab conjugated TNP-470 to HPMA to form caplostatin (Satchi-Fainaro et al., 2005). Caplostatin can be administered over a dose range more than 10-fold that of the original TNP-470 without any toxicity. In addition to its antiangiogenic activity, caplostatin is the most potent known inhibitor of vascular permeability (Satchi-Fainaro et al., 2005).

9.1.7 Thrombospondin-1

Thrombospondin-1 (TSP-1) was the first protein to be recognized as a naturally occurring inhibitor of angiogenesis (Good et al., 1990). TSP-1, a heparin-binding protein that is stored in extracellular matrix, was able to inhibit proliferation of endothelial cells from different tissues (Taraboletti et al., 1990) and appeared to destabilize contacts between endothelial cells (Iruela-Arispe et al., 1991). Tumors grew significantly faster in TSP-1 null mice than in wild-type mice (Lawler, 2002). Bocci et al. (2003) showed that the antiangiogenic chemotherapy increased circulating TPS-1, and that the deletion of TSP-1 in mice completely abrogated the antitumor effect of this antiangiogenic therapy.

The fact that TSP-1 is a potent endogenous inhibitor of angiogenesis prompted several groups to explore therapeutic applications of TSP-1, through the identification of strategies to upregulate endogenous TSP-1 and the delivery of recombinant TSP-1 repeats (TSRs) or synthetic peptides that contain sequences from TSRs.

9.1.8 Angiostatin and Endostatin

They were discovered by M. O'Reilly in Folkman's laboratory based on Folkman's hypothesis of a mechanism to explain the phenomenon that surgical removal of certain tumors leads to rapid growth of remote metastases. This hypothesis said that if tumors produce both stimulators and inhibitors of angiogenesis, an excess of inhibitors could accumulate within an angiogenic tumor. In the circulation, however, the ratio would be reversed. Angiogenesis inhibitors would increase relative to stimulators, because of the rapid clearance of stimulators from the blood. Folkman formulated this hypothesis after reading Bouck's first report in 1989 that the emergence of tumor angiogenesis was the result of a shift in balance between positive and negative regulators of angiogenesis in a tumor (Rastinejad et al., 1989). Bouck reported that the switch to angiogenesis during tumorigenesis of transformed hamster cells was associated with the downregulation of an inhibitor of angiogenesis, TSP. She suggested that the switch to the angiogenic phenotype could be the result of a shift in the net balance of positive and negative regulators of angiogenesis.

In 1991, O'Reilly began to screen a variety of transplantable murine tumors for their ability to suppress metastases. A Lewis lung carcinoma was the most efficient. When the metastasis-suppressing primary tumor was present in the dorsal subcutaneous position, microscopic lung metastases remained dormant at a diameter of less than 200 μm surrounding a pre-existing microvessel, but revealed no new vessels. Within 5 days after surgical removal of the primary tumor, lung metastases became highly angiogenic and grew rapidly, killing their host by 15 days (O'Reilly et al., 1994). This striking evidence that the primary tumor could suppress angiogenesis in its secondary metastases by a circulating inhibitor was further supported by the demonstration that a primary tumor could also suppress corneal angiogenesis by an implanted pellet of FGF-2. O'Reilly then succeeded in purifying this inhibitor from the serum and urine of tumor-bearing animals. It was a 38-kDa internal fragment identical in amino acid sequence to the first four kringle structures of plasminogen and it was named angiostatin. Angiostatin specifically inhibited the proliferation of growing vascular endothelial cells and had no effect on resting confluent endothelial cells or on other cell types, including SMCs, epithelial cells, fibroblasts, and tumor cells. It also inhibited growth of primary tumors by up to 98% (O'Reilly et al., 1996) and was able to induce regression of large tumors (1–2% of body weight) and maintain them at a microscopic dormant size.

Based upon the same rationale and strategy, O'Reilly isolated and purified another angiogenesis inhibitor from a murine hemangioendothelioma. This inhibitor, called endostatin (O'Reilly et al., 1997), is a 20-kDa protein with an N-terminal amino acid sequence identical to the carboxy terminus of collagen XVIII. It was purified directly from tumor cell-conditioned medium. Endostatin is also a specific inhibitor of endothelial proliferation and has no effects on resting endothelial cells or on other cell types. It is slightly more potent than angiostatin, and also causes regression of large tumors to a microscopic size. O'Reilly in the Folkman's lab found endostatin in the blood and urine of mice bearing tumors, which suppressed angiogenesis in remote metastases. In tumor-bearing animals continuous dosing of endostatin by an intraperitoneal mini-osmotic pump inhibited tumor growth 10-fold more effectively than the same dose administered once per day as a bolus dose (Kisker et al., 2001).

When endostatin is overexpressed in the vascular endothelium of mice tumors grow 300% more slowly in mice expressing only 1.6-fold more endostatin than wild-type mice (Sund et al., 2005). Recombinant endostatin was at first produced in *Escherichia coli*. Preparations of inclusion bodies that were endostatin-free and of low solubility were capable of regressing a variety of established murine tumors when administered subcutaneously (Boehm et al., 1997). When soluble recombinant endostatin was produced in yeasts, active endostatin was produced by numerous laboratories and a wide range of inhibited tumors were reported (Folkman and Jalluri, 2003).

More than 750 reports on endostatin reveal significant inhibition of more than 20 different rat and human tumors (in mice) by administration of the recombinant endostatin protein (Table 9.6). Endostatin counteracts virtually all the angiogenic genes upregulated by either VEGF or FGF-2 and also downregulates endothelial

Table 9.6 Human tumors in mice inhibited by recombinant endostatin protein

Ovarian
Colorectal carcinoma
Spontaneous mammary carcinoma
B-16 melanoma
Hepatoma
Lung adenocarcinoma
Lewis lung carcinoma
Human non-small cell carcinoma
Rat glioma
Human myeloid leukemia in SCID mice
Non-Hodgkin lymphoma
Pancreatic carcinoma
Laryngeal squamous cell carcinoma
Glioblastoma
Prostate carcinoma
Neuroblastoma
Testicular carcinoma
Breast carcinoma
Head and neck squamous cell carcinoma
Kaposi's sarcoma
Renal cell cancer
Bladder cancer

cell *Jun B*, HIF-1α, NRP, and the EGF receptor (Abdollahi et al., 2004). However, clinical trials using endostatin in cancer patients have been only sporadically positive.

It has been demonstrated that individuals with Down's syndrome have higher levels of circulating endostatin than normal individuals because of an extra-copy of the gene for the endostatin precursor on chromosome 21 (Zorick et al., 2001). Interestingly, these subjects have the lowest incidence of 200 human cancers compared with age-matched controls (Yang et al., 2002). Mice that were engineered to genetically overexpress endostatin to mimic individuals with Down's syndrome have tumors that grow 300% slower (Sund et al., 2005).

9.1.9 Thalidomide

In 1950s thalidomide was developed as a sedative that showed non-toxicity in pre-clinical animal models. In 1962, the association between limb defects in babies born to mothers who used thalidomide during pregnancy was established (Mellin and Katzenstein, 1962) (Table 9.7).

In 1994, D'Amato in the Folkman's laboratory reported that thalidomide is an angiogenic inhibitor (D'Amato et al., 1994). Corneal neovascularization in rabbits induced by FGF-2 or VEGF was blocked by orally administered thalidomide. This activity of thalidomide was mainly the result of its direct effect on inhibiting new blood vessel formation and not on suppression of infiltrating host inflammatory cells. Histologic sections of the pretreated neovascularized corneas were virtually

Table 9.7 Milestones in the clinical development of thalidomide

1954
Thalidomide is synthesized by Grunenthal
1956
Implementation as sedative in Germany
1961
Detection of birth defects and neuropathy caused by thalidomide. Subsequent withdrawal
1965
Proof of effectiveness in erythema nodosum leprosum (ENL)
1988
Implementation in graft-versus-host disease
1991
Confirmation of inhibition of lypopolysaccaridic-induced TNF-α expression
1994
Detection of the antiangiogenic properties of thalidomide
1996
Design of the first thalidomide analogs with anti-TNF-α effect (IMiDs)
2000
Discovery of effectiveness in multiple myeloma
2002
IMiDs in multiple myeloma
2005
Therapeutic effect in combination therapy and first-line treatment confirmed in randomized studies

free of inflammatory cells. Thalidomide also inhibited corneal neovascularization in mice, but it was necessary to give the drug through the intraperitoneal route and at high doses, because mice do not metabolize thalidomide effectively.

Thalidomide has been shown to have pleiotropic effects including antiangiogenic (downregulation of TNF-α, FGF-2, and VEGF) immunomodulatory, neurologic, and antiinflammatory effects (Ribatti and Vacca, 2005). Thalidomide was approved in Australia for the treatment of advanced multiple myeloma in 2003 and now is used as a first-line therapy. Many patients have been on the drug for 3–5 years without evidence of drug resistance (Ribatti and Vacca, 2005).

Soon after the discovery of the beneficial clinical effects of thalidomide for multiple myeloma and other tumors, a search of analogs with improved therapeutic efficacy and fewer adverse effect began. Synthetic analogs of thalidomide, also referred to as immunomodulatory drugs (IMiDs) have been developed to reduce the side effects of thalidomide and to enhance the antiinflammatory and anticancer effects. Indeed, IMiDs revealed significantly more antiangiogenic potential than thalidomide in in vitro assay (Dredge et al., 2002).

9.2 Metronomic Chemotherapy

T. Browder in the Folkman's lab was the first to demonstrate this novel concept: by optimizing the dosing schedule of conventional cytotoxic chemotherapy to achieve more sustained apoptosis of endothelial cells in the vascular bed of a tumor, it is possible to achieve more effective control of tumor growth in mice, even if the tumor cells are drug-resistant (Browder et al., 2000). Browder reported that conventional chemotherapy such as cyclophosphamide administered by the traditional schedule of maximum tolerated doses (MTD) interspersed with off-therapy intervals of 2–3 weeks to allow the bone marrow and gastrointestinal tract recover led to a drug resistance in all tumors when therapy was started in Lewis lung carcinomas at tumor volumes of 100–650 mm^3 (Browder et al., 2000).

They found that cyclophosphamide when administered at the MTD caused apoptosis of endothelial cells in the newly formed tumor microvessels. However, this antiangiogenic effect did not translate into significant therapeutic benefit because the damage to the vasculature was largely repaired during the long recovery periods between successive cycles of MTD-based therapy.

In contrast, antiangiogenic chemotherapy is administered more frequently at lower doses, without long interruptions in therapy, and with little or no toxicity and when cyclophosphamide was administered at more frequent intervals and at lower doses, it acted as an angiogenesis inhibitor. Proliferating endothelial cells in the tumor vascular bed underwent a wave of apoptosis around 4 days before tumor cell apoptosis began. All tumors completely regressed and animals remained tumor-free for their normal lifespan (up to 657 days).

Furthermore, Browder and co-workers administered a combination of the angiogenesis inhibitor TNP 470 with weekly doses of cyclophosphamide to treat transplanted mouse tumors that were previously selected for resistance to cyclophosphamide. They found that the combined treatment could cause marked and sustained regression of such tumors (Browder et al., 2000).

In an editorial, Hanahan coined the term "metronomic chemotherapy" to indicate the new schedule itself (Hanahan et al., 2000). During antiangiogenic chemotherapy, endothelial cell apoptosis and capillary dropout precede the death of tumor cells that surround each capillary (Browder et al., 2000). Cyclophosphamide, 5-fluorouracil, 6-mercaptopurine ribose phosphate, and Doxil (the pegylated liposomal formulation of doxorubicin) inhibit angiogenesis when administered on an antiangiogenic dose schedule. These results have been confirmed by others and have also been modified with daily oral administration of the drug through drinking water, which seems to be less toxic than the weekly regimen.

Kerbel showed that continuous administration of cyclophosphamide in the drinking water inhibited tumor growth in mice by 95% and significantly increased circulating levels of TSP-1 (Bocci et al., 2003). The low dose "metronomic" chemotherapy was ineffective in TSP-1 null mice, indicating that the low dose oral chemotherapy was in part dependent on its capacity to induce an increase in circulating TSP-1. So, "metronomic chemotherapy" might not necessarily act directly on endothelial cells, but might instead act by inducing endothelial cell-specific inhibitors, such as TSP-1. (Table 9.8).

Pediatric oncologists use a metronomic-like modality of chemotherapies called "maintenance chemotherapy" to treat various pediatric malignancies such as acute lymphoblastic leukemia, neuroblastoma, or Wilm's tumor (Kamen et al., 2006).

A combination of cytotoxic drugs (taxanes, cisplatin, or 5-fluorouracil) with angiogenesis inhibitors (TNP 470, endostatin, SU11248) produced at least additive, but in certain cases synergistic antitumoral effects (Gasparini et al., 2005).

Combinatorial therapies with antiangiogenic agents are not limited to those including cytotoxic chemotherapy. Several preclinical and clinical trials are exploring the combination of various angiogenesis inhibitors with other targeted therapies, such as EGFR or Ger2 inhibitors (cetuximab, erlotinib, trastuzumab), PDGFR/bcr-abl inhibitors (imatinib), proteasome inhibitors (bortezomib), and other antiangiogenic agents, such as inhibitors of integrins.

The major mechanisms accounting for the antitumor effects of metronomic chemotherapy regimes appear to be targeting the circulating bone marrow-derived endothelial progenitor cells and activated endothelial cells in the growing tumor neovasculature. In fact, "metronomic chemotherapy" can prevent mobilization of circulating endothelial progenitor cells and can act against not only the endothe-

Table 9.8 Possible mechanisms of the antiangiogenic basic of metronomic chemotherapy

Direct
Growth arrest and/or apoptosis of activated endothelial cells in tumor neovasculature
Sustained suppression of mobilization of endothelial progenitor cells

Indirect (through TSP-1)
Growth arrest and/or apoptosis of CD36$^+$ activated endothelial cells in tumor neovasculature
Sustained suppression of mobilization of endothelial progenitor cells

lial progenitors (Shaked et al., 2005), but also differentiated endothelial cells in the tumor neovasculature (Browder et al., 2000; Klement et al., 2000).

An important difference concerns the side effects of antiangiogenic therapy compared with chemotherapy. Bone marrow suppression, hair loss, severe vomiting and diarrhea, and weakness are less common with antiangiogenic therapy.

9.3 Receptor Tyrosine Kinase (RTK) Inhibitors

RTKs are transmembrane proteins containing an extracellular ligand-binding domain and an intracellular catalytic domain. Many of the processes involved in tumor growth, progression, and metastasis are mediated by signaling molecules acting downstream from activated RTK. In addition, several RTKs, most notably VEGFR and PDGFR are implicated in tumor-dependent angiogenesis. Therapeutic approaches using RTK inhibition have involved strategies to target either a single pathway or simultaneous inhibition of multiple pathways. The mechanism of action of all the RTK inhibitors consists of binding in the vicinity of the ATP binding site of their target tyrosine kinase thereby preventing phosphorylation of tyrosine residues of the receptor and subsequent intracellular signaling. Many RTK inhibitors have been developed which all target one or more specific receptors (Table 9.9).

A variety of small-molecules RTK inhibitors targeting the VEGF receptors have been described. Inhibitors of VEGF signaling not only stop angiogenesis but also cause regression of some tumor vessels (Bergers et al., 2003), causing changes in all components of the vessel wall of tumor, consisting in loss of endothelial fenestrations, regression of tumor vessels, and appearance of basement membrane ghosts

Table 9.9 Receptor tyrosine kinase (RTK) inhibitors and their targets

Agent	Targets
SU5416	VEGFR-1, VEGFR-2, KIT
SU6668	VEGFR-2, PDGFRβ, FGFR-1, KIT
SU11248	VEGFR-1, VEGFR-2, PDGFRα/β, KIT, FlT-3
SU14813	VEGFR-1, VEGFR-2, PDGFRα/β, KIT, FlT-3
PTK787/ZK222584	VEGFR-2, PDGFRβ, KIT
CP-547, 632	VEGFR-2, FGFR-1
AG013736	VEGFR-2, PDGFRβ
AZD2171	VEGFR-1, VEGFR-3
AMG706	VEGFR-1, VEGFR-3
CEP-7055	VEGFR-1, VEGFR-3, Flt-3, Mlkl-3
CEP-701	Flt-3, Trk kinases
PKC-412	PKC, VEGFR-2, PDGFR, c-kit, Flt-3
MLN-518	PDGFRβ, c-kit, Flt-3
GW-572016	EGFR, Her2
EKB-569	EGFR, Her2 (irreversible inhibitor)
PKI-166	EGFR, Her2
CI-1033	EGFR, Her2 (irreversible inhibitor)
OSI-774/erlotinib	EGFR
ZD1839/gefitinib	EGFR
ST1571/imatinib	Bcr/Abl, PDGFRβ, c-kit

(Inai et al., 2004). Moreover, targeting VEGF/VEGFR system transforms some tumor capillaries into a more normal phenotype (Jain et al., 2006). Finally, VEGF inhibition might have direct cytotoxic effects on tumor cells that aberrantly express VEGF receptors and depend to some extent on VEGF for their survival.

The most advanced are SU11248/sunitinib and BAT43-9006/Sorafenib. SU11248 inhibits VEGFR-1, VEGFR-2, PDGFR, c-kit, and Flt-3 (Smith et al., 2004) and received FDA approval in January 2006 for patients with gastrointestinal stromal tumors and advanced kidney cancer. BAT43-9006 was initially identified as a raf kinase inhibitor and subsequently shown to inhibit several RTKs including VEGFR. In September 2005, Sorafenib received FDA approval for the treatment of renal cell carcinoma. An additional VEGF RTK inhibitor in late-stage clinical trials is PKT 787/ZK222584, which inhibits VEGFR-2 and at higher concentrations PDGFR, c-kit, and c-Fms (Wood et al., 2000). Other anti-VEGF agents including VEGF-Trap (Regeneron), a soluble receptor targeting VEGF-A, VEGF-B, and PlGF, an anti-sense oligonucleotide VEGF-AS (Vasgene Therapeutics, Inc.) targeting VEGF-A, VEGF-C, and VEGF-D, are at various stages of clinical development (Jain et al., 2006). The VEGF-Trap abolished mature, pre-existing vasculature in established xenografts, resulting in almost completely avascular tumors subsequently followed by marked tumor regression and suppressed tumor growth (Holash et al., 2002). The VEGF-Trap has performed impressively in extensive animal studies of cancer and eye diseases, and clinical trials appear promising.

AZD2171 is one of the most potent inhibitors of VEGFR-2 and also of VEGFR-1 and VEGFR-3. Consistent with an antiangiogenic effect, one daily treatment with AZD2171 produced dose-dependent inhibition of tumor growth in a broad range of human tumor xenografts.

Strong preclinical evidence points to important crosstalk between activated EGFR and VEGFR pathways in tumor biology. In preclinical studies, resistance to anti-EGFR therapy could be circumvented and survival could be improved with antiangiogenic therapy. Therefore, the combination of inhibitors of both pathways may be more effective than blocking one or the other.

The RTK inhibitors of angiogenesis are generally well tolerated as an oral medication and they are relatively easy to deliver. It is likely that they will need to be given in a combination with chemotherapy or together with other biological-targeted agents.

9.4 Inhibitors of VEGF

VEGF is overexpressed in the majority of solid tumors. Inhibitors of VEGF signaling not only stop angiogenesis but also cause regression of some tumor vessels (Bergers et al., 2003), causing robust and rapid changes in all components of the vessel wall of tumor, consisting in loss of endothelial fenestrations, regression of tumor vessels, and appearance of basement membrane ghosts (Inai et al., 2004).

Owing to the fact that many of the abnormalities, such as increased permeability, vasodilation, and tortuosity, of the tumor vascular network are secondary to VEGF,

Jain in 2005 has hypothesized that VEGF-targeted therapy can "normalize" the vasculature of the tumors. Blockade of VEGF signaling prunes the immature and leaky vessels of transplanted tumors in mice and actively remodels the remaining vasculature. This "normalized" vasculature is characterized by less leaky, less dilated, and less tortuous vessels with a more normal basement membrane and greater coverage by pericytes (Jain et al., 2006). Moreover, VEGF-targeted therapy can improve blood flow with a consequently increased delivery of chemotherapy and oxygen.

VEGF inhibition might have direct cytotoxic effects on tumor cells that aberrantly express VEGFRs, as well as NRP-1 and NRP-2, and depend to some extent on VEGF for their survival. VEGF provides a survival signal for breast carcinoma cells in vitro and blockade of VEGF results in apoptosis of these cells (Bachelder et al., 2001). In vitro studies using VEGFR-1 expressing human colon cancer cells showed that a monoclonal antibody to VEGFR-1 blocked tumor migration and invasion (Fan et al., 2005). Wu et al. (2006) demonstrated that anti-VEGFR-1 monoclonal antibody could block intracellular signaling and growth of human breast cancer xenografts.

Another mechanism by which VEGF-targeted therapy might be effective, is its role in blocking chemotaxis of bone marrow-derived progenitor cells by cytotoxic agents. Mobilization and recruitment of bone marrow-derived endothelial precursor cells to tumor vasculature are rapidly induced by the administration of vascular disrupting agents, and is probably due to VEGF-induced chemotaxis (Mimura et al., 2007).

There are currently over 20 different VEGF-targeted agents in clinical trials. Monoclonal antibodies selectively target components of the VEGF pathway and this approach has less toxicity.

9.5 Antiangiogenic Monoclonal Antibodies: Toward the Discovery of Avastin

Since 1994, the FDA has approved 20 therapeutic monoclonal antibodies for clinical use in the USA, including 9 for oncology indications (Table 9.10). In 1993, Ferrara's laboratory reported that a mouse anti-human VEGF monoclonal antibody called A4.6.1 exerted a potent inhibitory effect on the growth of three tumor cell lines injected subcutaneously in nude mice, the G55 glioblastoma multiforme, the SK-LMS-1 leiomyosarcoma, and the A673 rhabdomyosarcoma, whereas the antibody had no effects on the tumor cell in vitro, consistent with the hypothesis that the inhibition of angiogenesis is the mechanism of tumor suppression in vivo (Kim et al., 1993). The growth inhibition ranged between 70% and more than 95%. These findings provided the first direct demonstration that inhibition of the action of an endogenous endothelial cell mitogen may result in suppression of tumor growth in vivo.

Subsequently, Ferrara's laboratory demonstrated that many other cell lines were inhibited in vivo by anti-VEGF monoclonal antibodies (Warren et al., 1995;

Table 9.10 FDA approved monoclonal antibody therapeutics in oncology

Year approved	Antibody	Target	Indication
1977	Rituxan (rituximab)	CD20	NHL
1998	Herceptin (trastuzumab)	HER2	Breast cancer
2000	Mylotarg (Gemtuzumab ozogamicin)	CD33	AML
2001	Campath-1H (alentuzumab)	CD52	B cell CLL
2002	Zevalin (ibritumomab tiuxetan)	CD20	NHL
2003	Bexxar (tositumomab)	CD20	NHL
2004	Erbitux (cetuximab)	EGFR	colorectal and head and neck cancer
2004	Avastin (bevacizumab)	VEGF	colorectal and NSCLC
2006	Vectibix	EGFR	colorectal cancer

Borgstrom et al., 1996; 1998; 1999; Mesiano et al., 1998). The density of blood vessels was significantly lower in sections of tumors from antibody-treated animals as compared with controls. Furthermore, neither the antibodies nor VEGF had any effects on the in vitro growth of tumor cells. Intravital videomicroscopy techniques have allowed a more direct verification of the hypothesis that anti-VEGF antibodies indeed block tumor angiogenesis (Borgstrom et al., 1996). Tumor spheroids of A673 cells were implanted in dorsal skinfold chambers inserted in nude mice. Non-invasive imaging of the vasculature revealed a nearly complete suppression of tumor angiogenesis in anti-VEGF-treated animals as compared with controls. Histological analysis showed a dramatic difference in the density of CD34-positive vascular elements between the two groups. Warren et al. (1995) demonstrated that VEGF is a mediator of the in vivo growth of human colon carcinoma HM7 cells in a orthotopic nude mouse model of liver metastasis. In this murine model the expression of Flk-1 mRNA was markedly upregulated in the vasculature associated with liver metastases. Treatment with anti-VEGF monoclonal antibodies resulted in a dramatic decrease in the number and size of metastases. Also, neither blood vessels nor VEGFR-2 mRNA expression could be demonstrated in such metastases.

Three different human tumor cell lines (UB7, P-MEL, and LS174T) were implanted in two locations in immunodeficient mice, the cranium and the dorsal skinfold (Yuan et al., 1996). Treatment with an anti-VEGF monoclonal antibody was initiated when the tumor xenografts were already established and vascularized and resulted in time-dependent reductions in vascular permeability. These effects were accompanied by striking changes in the morphology of the vessels with dramatic reduction in diameter and tortuosity (Yuan et al., 1996).

The first antiangiogenic agent approved by the FDA was bevacizumab (Avastin, Genentech, Inc.), a humanized version (Presta et al., 1997) of the mouse anti-VEGF monoclonal antibody A4.6.1 used in early proof of concept studies (Kim et al., 1993). By site-directed mutagenesis of a human antibody framework, the residues involved in the six complementary-determining regions, and also several framework residues were changed to murine counterparts. Like its murine counterpart, bevacizumab binds and neutralizes all human VEGF-A isoforms and bioactive proteolytic fragments, but not mouse or rat VEGF. Bevacizumab does not neutralize other members of the VEGF gene family, such as VEGF-B or VEGF-C.

Bevacizumab inhibited the growth of human tumor cell lines in nude mice, achieving a maximal inhibition at the dose of 1–2 mg/kg twice weekly (Presta et al., 1997). The magnitude of the inhibition was inversely related to the content of stromal-derived mouse VEGF within the tumor xenograft. In tumors with high human/mouse VEGF ratio, the inhibition can exceed 90% (Warren et al., 1995; Gerber et al., 2000).

In 1997, Genentech initiated Phase I clinical trials with bevacizumab, showing that the antibody as a single agent was relatively non-toxic and that adding it to standard chemotherapy regimen did not significantly exacerbate chemotherapy-associated toxicities (Margolin et al., 2001; Gordon et al., 2001). In 1998, several Phase II studies were initiated with bevacizumab in different tumor types, either as single agent or in combination with chemotherapy. The most encouraging efficacy results were seen when bevacizumab was combined with standard first-line chemotherapy in metastatic colorectal cancer (Kabbinavar et al., 2003), and when was used as a single agent in renal cell cancer (Yang et al., 2003).

The clinical trial that resulted in FDA approval of Avastin in February 2004 was a randomized double-blind phase III study in which Avastin was administered in combination with bolus IFL (irinotecan, 5FU, leucovirin) chemotherapy as first-line therapy for previous untreated metastatic colorectal cancer (Hurwitz et al., 2004). Median survival was increased from 15.6 months in the bolus-IFL + placebo arm to 20.3 months in the bolus-IFL + bevacizumab arm. Similar increases were seen in progression-free survival, response rate, and duration of response. The clinical benefit of Avastin was seen in all subject subgroups (Hurwitz et al., 2004). Hypertension was more common in the IFL/Avastin-treated group, but was readily managed in all cases with oral antihypertensive agents (Hurwitz et al., 2004).

Treatment with bevacizumab is generally safe and well tolerated, but it can be accompanied by a variety of adverse effects (Table 9.11) which are broadened or intensified by concurrent chemotherapeutic agents (Hurvitz and Saini, 2006). Bevacizumab-related side effects are generally manageable; however, their monitoring is advised especially in patients with predisposing factors.

Table 9.11 Serious toxicities associated with bevacizumab

Bleeding
Proteinuria
Hypertension
Gastrointestinal perforation
Poor wound healing
Arterial thrombosis
Reversible posterior leukoencephalopathy syndrome

9.6 Microvascular Density Has Not Been Shown to Be a Valid Measurement to Guide or Evaluate Antiangiogenic Treatment

It is widely assumed that tumors with high microvascular density are good candidates for clinical trials of antiangiogenic therapies, whereas tumors that typically have low microvascular density are thought to be poor candidates for such clinical trials (Lenz, 2005; Rhee and Hoff, 2005; Zhong and Bowen, 2006; Cooney et al., 2006; Laquente et al., 2007). However, experimental evidence shows that both poorly vascularized and highly vascularized tumors can respond to antiangiogenic therapy. Microvascular density, accordingly offers no indication as to which patients might be most responsive to antiangiogenic therapy (Lenz, 2005; Rhee and Hoff, 2005; Zhong and Bowen, 2006; Cooney et al., 2006; Laquente et al., 2007). In addition, although a decrease in microvascular density following antiangiogenic therapy can give an indication of the antivascular activity of a particular agent, microvascular density as a single end point fails to provide an adequate measurement for resolving the vascular response to antiangiogenic agents.

Measuring a slight decrease, no change, or even an increase in microvascular density is still consistent with a vessel inhibition, because microvascular density is a dynamically complex quantity that is influenced by the initial vascular suppression plus the consequent interaction between the vascular and tumor-cell compartments (Hlatky et al., 2002).

During tumor regression induced by an angiogenesis inhibitor, microvascular density may decrease if capillary dropout exceeds tumor cell dropout, increase if tumor cell dropout exceeds capillary dropout, or remain the same if disappearance of capillaries and tumor cells parallel each other. Therefore, the detection of a decrease in microvascular density during treatment with an angiogenesis inhibitor, suggests that the agent is active. However, the absence of a decrease in microvascular density does not indicate that the agent is ineffective.

Singhal et al. (1999) showed that multiple myeloma that was resistant to high-dose chemotherapy regressed in response to treatment with a single antiangiogenic agent thalidomide, even though not all tumor regressions were accompanied by a decrease in microvascular density. The observation of tumor regression without a corresponding decrease in microvascular density does not indicate that mechanisms other than antiangiogenesis play a causal role in treatment response. The tumor cell population may simply decrease in direct proportion to, and as a direct consequence of, the loss of its supporting vasculature. Likewise, tumors undergoing antiangiogenic intervention may also follow a "shrink to fit" adaptation, which as a result may not lead to reduced microvascular density counts. It has consequently been argued that microvascular density reduction may not be the appropriate and expected readout of the success of antiangiogenic interventions in human tumors (Hlatky et al., 2002).

Changes in microvascular density do not independently measure vascular inhibition, but rather, reflect the changing ratio of the vascular component of the tumor to its tumor-cell component. Under antiangiogenic therapy, capillary inhibition or elimination occur first, followed by tumor-cell elimination, and both influence microvascular density.

Moreover, experimental antiangiogenic experiments in murine tumor models usually have a significant reduction of the microvascular density as a primary experimental readout of an antiangiogenic intervention. Human tumors have different growth kinetics compared with experimental tumors.

9.7 Perspectives

Some general questions remained unanswered and emerged. They include the identification for surrogate markers of antiangiogenesis, the understanding how antiangiogenic therapy and chemotherapy synergize, the characterization of the biological consequences of sustained suppression of angiogenesis on tumor biology and normal tissue homeostasis, and the mechanism of tumor escape from antiangiogenesis.

As concerns the identification of markers of angiogenesis, the levels of various tumor-produced angiogenic cytokines, including VEGF and FGF-2, in the circulation and body fluids of cancer patients have been used as indicators of the progression of malignant diseases. However, VEGF and FGF-2 have short half-lives in the ranges of minutes in the circulation, which make it very difficult to estimate the actual expression of these growth factors in tumors. There is an urgent need to identify biomarkers that allow us to select potentially responsive patients, to design the therapeutic protocols, and to monitor the therapeutic efficacies.

Approaches based on genomics and proteomics have provided insights into the mechanistics processes of angiogenesis and facilitated the identification of novel candidate therapeutic targets and allowed the monitoring of drug activities (Mittal and Nolan, 2007).

There are currently over 1,000 clinical trials investigating over 40 antiangiogenic agents in cancer treatment (www.clinicaltrials.gov). However, the results from these clinical trials have not shown the dramatic antitumor effects which were expected following preclinical studies. It has always been a challenge to extrapolate animal data into the clinical settings or this may be because of inadequate trial design in earlier studies.

The main problem in the development of antiangiogenic agents is that multiple angiogenic molecules may be produced by tumors, and tumors at different stages of development may depend on different factors for their blood supply. Although VEGF is expressed by up to 60% of human tumors, most tumors express five to eight other angiogenic proteins, for example, human breast cancer can express up to six angiogenic proteins. An analysis of human breast cancer biopsies covering a spectrum from low to high-grade malignancies revealed that late-stage breast cancers expressed a plethora of angiogenic factors, in contrast to earlier stage lesions, which preferentially expressed VEGF (Relf et al., 1997). Therefore, blocking a single angiogenic molecule was expected to have little or no impact on tumor growth.

When the expression of one angiogenic protein is suppressed for a long period, the expression of other angiogenic proteins might emerge (Dorrell et al., 2007). However, in apparent contrast with this view, experiments with neutralizing antibodies and other inhibitors demonstrated that blockade of VEGF alone can substantially suppress tumor growth and angiogenesis in several models. However, anti-VEGF monotherapy might even help tumors to switch on VEGF-unrelated angiogenic pathways for growing blood vessels (Ferrara and Kerbel, 2005). Consistent with this notion, tumors have been reported to escape from anti-VEGF therapy after a relative long-term treatment in a transgenic mouse tumor model (Casanovas et al., 2005).

Almost all clinical trials for evaluation of therapeutic efficacy of antiangiogenic agents are conducted in patients with advanced stage cancer and metastases. Currently, the majority of FDA-approved angiogenesis inhibitors, as well as those in Phase III clinical trials, neutralize VEGF, target its receptor, or suppress its expression by tumor cells. The most successful approaches are likely to involve combinatorial strategies to target cancer cells themselves along with perivascular and inflammatory cells.

A growing amount of evidence indicates that tumors react to therapy by upregulating angiogenic factors and mobilizing bone marrow-derived circulating endothelial progenitor cells (Bocci et al., 2004; Shaked et al., 2006). Patients treated with antiangiogenic drugs who show an initial antitumor response eventually show disease progression, suggesting that tumor develops resistance to this class of drugs (Hurwitz et al., 2004) (Table 9.12). The mechanisms responsible for resistance to antiangiogenic drugs are likely to be complex and are yet to be defined. Possible mechanisms for acquired resistance to antiangiogenic drugs (Kerbel et al., 2001; Sweeney et al., 2003) include selection and overgrowth of tumor variants that are hypoxia-resistant and thus less dependent on angiogenesis (Yu et al., 2002) and tumor-vessel remodeling resulting in a shift to mature, stabilized vessels that are less responsive to antiangiogenic drugs (Gladebender et al., 2004).

Selection of optimal drug combinations, cancer types, dosages, and patients is essential for designing successful clinical trials for antiangiogenic therapy. An ideal angiogenesis inhibitor should be orally bioavailable with acceptable short-term and long-term toxicities and have a clinically useful antitumor effect.

Polyphenols in green tea was one of the first oral angiogenesis inhibitors discovered to counteract VEGF-A-induced angiogenic activity (Cao and Cao, 1999). Epigallo-catechin-3-gallate when administered orally is able to inhibit mouse corneal angiogenesis and VEGF production (Cao and Cao, 1999).

Table 9.12 Mechanisms of resistance to angiogenesis inhibition

Endothelial cell heterogeneity
Angiogenic factor heterogeneity
Tumor cell heterogeneity
Impact of the tumor microenvironment
Compensatory responses to treatment
Angiogenesis independent tumor growth
Pharmacokinetics resistance

Table 9.13 Toxic effects of angiogenesis inhibitors

Bleeding, disturbed wound healing
Thrombotic events
Hypertension
Hypothyroidism
Fatigue
Proteinuria and edema
Leukopenia, lymphopenia, and immunomodulation
Dizziness, nausea, vomiting, and diarrhea
Skin toxicity including rash and hand–foot syndrome

Although most patients tolerate antiangiogenic therapy well, rare but severe toxicities, including hypertension, cardiovascular thrombosis, gastrointestinal perforation, delayed wound healing, and hemorrhages, have been observed (Table 9.13). Moreover, carefully constructed clinical trials with valid end points need to be executed. Finally, cancer genomics and proteomics are likely to identify novel tumor-specific endothelial targets and accelerate drug discovery.

References

Abdollahi A, Hahnfeldt P, Maercker C et al (2004) Endostatin's antiangiogenic signaling network. Mol Cell 13:649–663

Abramsson A, Lindblom P, Betsholtz C (2003) Endothelial and nonendothelial sources of PDGF-B regulate pericyte recruitment and influence vascular pattern formation in tumors. J Clin Invest 112:1142–1151

Achilles EG, Fernandez A, Allred EN et al (2001) Heterogeneity of angiogenic activity in a human liposarcoma: a proposed mechanism for 'no take' of human tumors in mice. J Natl Cancer Inst 93:1075–1081

Ahmad SA, Liu W, Jung YD et al (2001) The effects of angiopoietin-1 and -2 on tumor growth and angiogenesis in human colon cancer. Cancer Res 61:1255–1259

Aguayo AR, Estey E, Kantarajian T et al (1999) Cellular vascular endothelial growth factor is a predictor of outcome in patients with acute myeloid leukemia. Blood 94:3717–3721

Aguayo AR, Kantarajian T, Manshouri T et al (2000) Angiogenesis in acute and chronic leukemia and myelodysplastic syndromes. Blood 96:2240–2245

Aicher A, Heeschen C, Midner-Rihm C et al (2003) Essential role of endothelial nitric oxide synthase for mobilization of stem and progenitor cells. Nat Med 9:1370–1376

Alexandrakis MG, Passam FH, Dambaki C et al (2004) The relation between bone marrow angiogenesis and the proliferation index Ki-67 in multiple myeloma. J Clin Pathol 57:856–860

Algire GH, Chalkley HW (1945) Vascular reactions of normal and malignant tissue *in vivo*. J Natl Cancer Inst 6:73–85

Almong N, Henke V, Flores L et al (2006) Prolonged dormancy of human liposarcoma is associated with impaired tumor angiogenesis. FASEB J 20:947–949

Angiolillo AL, Sgadari C, Taub DD et al (1995) Human interferon-inducible protein 10 is a potent inhibitor of angiogenesis in vivo. J Exp Med 182:155–162

Antonelli-Orlidge A, Saunders KB, Smith SR et al (1989) An activated form of transforming growth factor-β is produced by cocultures of endothelial cells and pericytes. Proc Natl Acad Sci USA 86:4544–4548

Arbeit J, Munger K, Howley P, et al (1994) Progressive squamous epithelial neoplasia in K14-HPV16 transgenic mice. J Virol 68: 4358–4368

Arbeit J, Olson D, Hanahan D (1996) Upregulation of fibroblast growth factors and their receptors during multistage epidermal carcinogenesis in K14-HPV16 transgenic mice. Oncogene 13:1847–1857

Asahara T, Murohara T, Sullivan A et al (1997) Isolation of putative progenitor endothelial cells for angiogenesis. Science 275:964–967

Asahara T, Masuda H, Takahashi T et al (1999a) Bone marrow origin of endothelial progenitor cells responsible for postnatal vasculogenesis in physiological and pathological neovascularization. Circ Res 85:221–228

Asahara T, Takahashi T, Masuda H et al (1999b) VEGF contributes to postnatal neovascularization by mobilizing bone-marrow derived endothelial progenitor cells. EMBO J 18:3964–3972

Asellius G (1627) De lactibus sive lacteis venis. Milan, Mediolani

Asosingh K, De Raeve H, Menu E, et al (2004) Angiogenic switch during 5T2MM murine myeloma tumorigenesis: role of CD45 heterogeneity. Blood 103:3131–3137

Auerbach R, Kubai L, Knighton D et al (1974) A simple procedure for the long-term cultivation of chicken embryos. Dev Biol 41:391–394

Ausprunk DH, Knighton DR, Folkman J (1974) Differentiation of vascular endothelium in the chick chorioallantois: a structural and autoradiographic study. Dev Biol 38:237–248

Ausprunk DH, Knighton DR, Folkman J (1975) Vascularization of normal and neoplastic tissues grafted to the chick chorioallantois. Role of host and preexisting graft blood vessels. Am J Pathol 79:597–618

Bachelder RE, Crago A, Chung J et al (2001) Vascular endothelial growth factor is an autocrine survival factor for neuropilin-expressing breast carcinoma cells. Cancer Res 61:5736–5740

Baird A, Esch F, Mormede P et al (1986) Molecular characterization of fibroblast growth factor: distribution and biological activities in various tissues. Recent Prog Horm Res 42:143–205

Balkwill F, Mantovani A (2001) Inflammation and cancer: back to Virchow? Lancet 357:539–545

Baluk P, Morikawa S, Haskell A et al (2003) Abnormalities of basement membrane on blood vessels and endothelial sprouts in tumors. Am J Pathol 163:1801–1815

Baron JA, Monzon F, Galaria N et al (2000) Angiomatoid melanoma: a novel pattern of differentiation in invasive periocular desmoplastic malignant melanoma. Hum Pathol 31:1520–1522

Bartlett MR, Underwood PA, Parish CR (1995) Comparative analysis of the ability of leucocytes, endothelial cells and platelets to degrade the subendothelial basement membrane: evidence for cytokine dependence and detection of a novel sulphatase. Immunol Cell Biol 73:113–124

Bauovois B, Dumont J, Mathiot C, et al (2002) Production of matrix metalloproteinase-9 in early C-CLL: suppression by interferons. Leukemia 16:791–798

Beck L Jr, D'Amore PA (1997) Vascular development: cellular and molecular regulation. FASEB J 11:365–373

Bellamy WT, Richter L, Sirjani D, et al (2001) Vascular endothelial growth factor (VEGF) is an autocrine promoter of abnormal localized immature precursors (ALIP) and leukemia progenitor formation in myelodysplastic syndromes. Blood 97:1427–1434

Bellman S, Odén B (1959) Regeneration of surgical divided lymph vessels. An experimental study on the rabbit's ear. Acta Chir Scand 116:99–117

Bellocq A, Antonie M, Flahault A et al (1998) Neutrophil alveolitis in bronchioalveolar carcinoma: induction by tumor-derived interlukin-8 and relation to clinical outcome. Am J Pathol 152:83–92

Benelli R, Morini M, Carrozzino F et al (2002) Neutrophils as a key cellular target for angiostatin: implications for regulation of angiogenesis and inflammation. FASEB J 16:267–269

Benitez-Bribiesca L, Wong A et al (2001) The role of mast cell tryptase in neoangiogenesis of premalignant and malignant lesions of the uterine cervix. J Histochem Cytochem 49:1061–1062

Benjamin LE, Keshet E (1997) Conditional switching of vascular endothelial growth factor (VEGF) expression in tumors: induction of endothelial cell shedding and regression of hemangioblastoma-like vessels by VEGF withdrawal. Proc Natl Acad Sci USA 94:8761–8766

Benjamin LE, Golijanin D, Itin A et al (1999) Selective ablation of immature blood vessels in established human tumors follows vascular endothelial growth factor withdrawal. J Clin Invest 103:159–165

Bergers G, Brekken R, McMahon G et al (2000) Matrix metalloproteinase-9 triggers the angiogenic switch during carcinogenesis. Nat Cell Biol 2:737–744

Bergers G, Song S, Meyer-Morse N et al (2003) Benefits of targeting both pericytes and endothelial cells in the tumor vasculature with kinase inhibitors. J Clin Invest 111:1287–1295

Betsholtz C. (2004) Insight into physiological functions of PGF through genetic studies. Cytokine Growth Factor Rev 15:215–228

Bingle L, Brown NJ, Lewis CE (2002) The role of tumour-associated macrophages in tumour progression: implications for new anticancer therapies. J Pathol 196:254–265

Birnbaum D, deLapeyriere O, Adnane J et al (1991) Role of FGFs and FGF receptors in human carcinogenesis. Ann NY Acad Sci 638:409–411

Bocci G, Francia G, Man S et al (2003) Thrombospondin-1, a mediator of the antiangiogenic effects of low-dose metronomic chemotherapy. Proc Natl Acad Sci USA 100:12917–12922

Bocci G, Man S, Green SK et al (2004) Increased plasma vascular endothelial growth factor (VEGF) as a surrogate marker for optimal therapeutic dosing of VEGF receptor-2 monoclonal antibodies. Cancer Res 64:6616–6625

Boehm T, Folkman J, Browder T et al (1997) Antiangiogenic therapy of experimental cancer does not induce acquired drug resistance. Nature 390:404–407

Borgstrom P, Hillan KJ, Sriramarao P et al (1996) Complete inhibition of angiogenesis and growth of microtumors by anti-vascular endothelial growth factor neutralizing antibody: novel concept of angiostatic therapy from intravital videomicroscopy. Cancer Res 56:4032–4039

Borgstrom P, Bourdon MA, Hillan KJ et al (1998) Neutralizing anti-vascular endothelial growth factor antibody completely inhibits angiogenesis and growth of human prostate carcinoma microtumors *in vivo*. Prostate 35:1–10

Borgstrom P, Gold DP, Hillan KJ et al (1999) Importance of VEGF for breast cancer angiogenesis *in vivo*: implications from intravital microscopy of combination treatments with an anti-VEGF neutralizing antibody and doxorubicin. Anticancer Res 19:4203–4214

Bornstein P (1992) Thrombospondins: structure and regulation of expression. FASEB J 6:3290–3299

Bossi P, Viale G, Lee Ak et al (1995) Angiogenesis in colorectal tumors: microvessel quantitation in adenomas and carcinomas with clinicopathological correlations. Cancer Res 55:5049–5053

Bossy-Wtzel E, Bravo R, Hanahan D (1992) Transcription factors JunB and cJun are selectively up-regulated and functionally implicated in fibrosarcoma development. Genes Dev 6:2340–2351

Bouck N, Stellmach V, Hsu S (1996) How tumors become angiogenic. Adv Cancer Res 69:135–174

Bowrey PF, King J, Magarey C et al (2000) Histamine, mast cell and tumor cell proliferation in breast cancer: does preoperative cimetidine administration have an effect? Br J Cancer 82:167–170

Boyle RG, Travers S (2006) Hypoxia: targeting the tumour. Anticancer Agents Med Chem 6:281–286

Brahimi-Horn C, Pouyssegur J (2006) The role of the hypoxia-inducible factor in tumor metabolism growth and invasion. Bull Cancer 93:E73–E80

Breiteneder-Geleff S, Soleiman A, Kowalski H et al (1999) Angiosarcoma express mixed endothelial phenotypes of blood and lymphatic capillaries: podoplanin as a specific marker for lymphatic endothelium. Am J Pathol 154:385–394

Brem S, Cotran R, Folkman J (1972) Tumor angiogenesis: a quantitative method for histologic grading. J Natl Cancer Inst 48:347–356

Brem S, Folkman J (1975) Inhibition of tumor angiogenesis mediated by cartilage. J Exp Med 141:427–439

Brem S, Brem H, Folkman J et al (1976) Prolonged tumor dormancy by prevention of neovascularization in the vitreous. Cancer Res 36:2807–2812

Brem S, Gullino PM, Medina D (1977) Angiogenesis: a marker for neoplastic transformation of mammary papillary hyperplasia. Science 195:880–882

Brem S, Jensen HM, Gullino PM (1978) Angiogenesis as a marker of preneoplastic lesions of the human breast. Cancer 41:239–244

Brouty-Boye D, Zetter BR (1980) Inhibition of cell motility by interferon. Science 208:16–18

Browder T, Butterfield CE, Kraling BM et al (2000) Antiangiogenic scheduling of chemotherapy improves efficacy against experimental drug-resistant cancer. Cancer Res 60:1878–1886

Brown LF, Van De Water L, Harvey VS et al (1988) Fibrinogen influx and accumulation of cross-linked fibrin in healing wounds and in tumor stroma. Am J Pathol 130:455–465

Brown LF, Yeo KT, Berse B et al (1992) Expression of vascular permeability factor (vascular endothelial growth factor) by epidermal keratinocytes during wound healing. J Exp Med 176:1375–1379

Burke B, Tang N, Corke KP et al (2002) Expression of HIF-1 α by human macrophages: implications for the use of macrophages in hypoxia-regulated cancer gene therapy. J Pathol 196:204–212

Busam KJ, Berwick M, Blession K et al (2005) Tumor vascularity is not a prognostic factor for malignant melanoma of the skin. Am J Pathol 147:1049–1056

Butler TP, Gullino PM (1975) Quantitation of cell shedding into efferent blood of mammary adeno-carcinoma. Cancer Res 35:512–516

Cantu De Leon D, Lopez-Graniel C, Frias Mendivil M et al (2003) Significance of microvascular density (MVD) in cervical cancer recurrence. Int J Gynecol Cancer 13:856–862

Cao Y, Cao R (1999) Angiogenesis inhibited by drinking tea. Nature 398:381

Cao Y (2005) Emerging mechanisms of tumor lymphangiogenesis and lymphatic metastasis. Nat Rev Cancer 7:192–198.

Carmeliet P, Ferreira V, Breier G et al (1996) Abnormal blood vessel development and lethality in embryos lacking a single VEGF allele. Nature 380:435–439

Carmeliet P, Moons L, Luttun A et al (2001) Synergism between vascular endothelial growth factor and placental growth factor contributes to angiogenesis and plasma extravasation in pathological conditions. Nat Med 7:575–583

Casanovas O, Hicklin DJ, Bergers G et al (2005) Drug resistance by evasion of antian-giogenic targeting of VEGF signaling in late-stage pancreatic islet tumors. Cancer Cell 8:299–309

Cavallo F, Quaglino E, Cifaldi L et al (2001) Interleukin-12 activate lymphocytes influence tumor genetic program. Cancer Research 61:3518–3523

Cavallo T, Sade R, Folkman J et al (1972) Tumor angiogenesis: rapid induction of endothelial mitosis demonstrated by autoradiography. J Cell Biol 54:408–420

Cavallo T, Sade R, Folkman J et al (1973) Ultrastructural autoradiographic studies of the early vaso-proliferative response in tumor angiogenesis. Am J Pathol 70:345–362

Cernea CR, Ferraz AR, de Castro IV et al (2004) Angiogenesis and skin carcinomas with skull base invasion: a case-control study. Head Neck 26:396–400

Chae SS, Paik JH, Furneaux H et al (2004) Requirement for sphingosine 1-phosphate receptor-1 in tumor angiogenesis demonstrated by in vivo RNA interference. J Clin Invest 114:1082–1089

Chalkley HW (1943) Method for the quantitative morphologic analysis of tissues. J Natl Cancer Inst 4:47–53

Chandrachud KM, Pendleton N, Chisholm DM et al (1997) Relationship between vascularity, age and survival in non-small-cell lung cancer. Br J Cancer 76:1367–1375

Chang YS, di Tomaso E, Mc Donald DM et al (2000) Mosaic blood vessels in tumors: frequency of cancer cells in contact with flowing blood. Proc Natl Acad Sci USA 97:14608–14613

Chantrain CF, Shimada H, Jodele S et al (2004) Stromal matrix metalloproteinase-9 regulates the vascular architecture in neuroblastoma by promoting pericyte recruitment. Cancer Res 64:1675–1686

Chavakis T, Cines DB, Rhee J-S et al (2004) Regulation of neovascularization by human neu-trophil peptides (α-defensins): a link between inflammation and angiogenesis. FASEB J 18:1306–1308

Christofori G, Naik P, Hanahan D (1995) Vascular endothelial growth factor and its receptors, flt-1 and flk-1 are expressed in normal pancreatic islets and throughout islet tumorigenesis. Mol Endocrinol 9:1760–1770

Clarijs R, Ruiter DJ, De Wall RM (2001) Lymphangiogenesis in malignant tumours: does it occur? J Pathol 193:143–146

Clark ER, Hitschler WJ, Kirby-Smith HT et al (1931) General observations on the ingrowth of new blood vessels into standardized chambers in the rabbit's ear, and the subsequent changes in the newly grown vessels over a period of months. Anat Rec 50:129–167

Clark ER, Clark EL (1932) Observations on the new growth of lymphatic vessels as seen in transparent chambers introduced into the rabbit's ear. Am J Anat 51:49–87

Clark ER, Clark EI (1939) Microscopic observations on the growth of blood capillaries in the living mammals. Am J Anat 64:251–301

Collins PD, Connolly DT, Williams TJ (1993) Characterization of the increase in vascular permeability induced by vascular permeability factor in vivo. Br J Pharmacol 109:195–199

Colombo MP, Trinchieri G (2002) Interleukin-12 in anti-tumor immunity and immunotherapy. Cytokine Growth Factor Rev 13:155–168

Colorado PC, Torre A, Kamphaus G et al (2000) Anti-angiogenic cues from vascular basement membrane collagen. Cancer Res 60:2520–2536

Condeelis J, Pollard JW (2006) Macrophages: obligate partners for tumor cell migration, invasion, and metastasis. Cell 124:263–266

Connolly DT, Huevelman DM, Nelson R et al (1989) Tumor vascular permeability factors stimulate endothelial cell growth and angiogenesis. J Clin Invest 84: 1470–1478

Conti P, Pang X, Boucher W et al (1997) Impact of Rantes and MCP-1 chemokines on in vivo basophilic cell recruitment in rat skin injection model and their role in modifying the protein and mRNA levels for histidine decarboxylase. Blood 89: 4120–4127

Cooney MM, van Heeckeren W, Bhakta S et al (2006) Drug insight: vascular disrupting agents and angiogenesis-novel approaches for drug delivery. Nat Clin Pract Oncol 3:682–692

Coussens LM, Web Z (1996) Matrix metalloproteinases and the development of cancer. Chem Biol 3:895–904

Coussens LM, Hanahan D, Arbeit J (1996) Genetic predisposition and parameters of malignant progression in K14-HPV16 transgenic mice. Am J Pathol 149:1899–1917

Coussens L, Raymond W, Bergers G et al (1999) Inflammatory mast cells up-regulate angiogenesis during squamous epithelial carcinogenesis. Genes Dev 13:1382–1397

Coussens L, Tinkle C, Hanahan D et al (2000) MMP-9 supplied by bone marrow-derived cells contribute to skin carcinogenesis. Cell 103: 481–490

Couvelard A, O'Toole D, Turley H et al (2005) Microvascular density and hypoxia-inducible factor pathway in pancreatic endocrine tumours: negative correlation of microvascular density and VEGF expression with tumour progression. Br J Cancer 92:94–101

Crivellato E, Nico B, Vacca A et al (2002) Mast cell heterogeneity in B cell non-Hodgkin's lymphomas: an ultrastructural study. Leuk Lymphoma 43:2201–2205

Crivellato E, Nico B, Vacca A, et al (2003) Ultrastructural analysis of mast cell recovery after secretion by piecemeal degranulation in B cell non-Hodgkin's lymphoma. Leuk Lymphoma 44:517–521

Cross MJ, Claesson-Welsch L (2001) FGF and VEGF function in angiogenesis: signalling pathways, biological responses and therapeutic inhibition. Trends Pharmacol Sci 22:201–207

Cursiefen C, Chen L, Borges LP et al (2004) VEGF-A stimulates lymphangiogenesis and hemangiogenesis in inflammatory neovascularization via macrophage recruitment. J Clin Invest 113:1040–1050

Daddras SS, Paul T, Bertoncini J et al (2003) Tumor lymphangiogenesis: a novel prognostic indicator for cutaneous melanoma metastasis and survival. Am J Pathol 162:1951–1960

Dales JP, Garcia S, Carpentier S et al (2004a) Long-term prognostic significance of neoangiogenesis in breast carcinomas: comparison of Tie-2/Tek, CD105, and CD31 immunocytochemical expression. Hum Pathol 35:176–183

Dales JP, Garcia S, Carpentier S et al (2004b) Prediction of metastasis risk (11 year follow-up) using VEGF-R1, VEGF-R2, Tie-2/Tek and CD105 expression in breast cancer (n = 905). Br J Cancer 90:1216–1221

Daldrup H, Shames DM, Wendland M et al (1998) Correlation of dynamic contrast-enhanced MR imaging with histologic tumor grade: comparison of macromolecular and small-molecular contrast media. Am J Roentgenol 171:941–949

Dameron KM, Volpert OV, Tainsky MA et al (1994) The p53 tumor suppressor gene inhibits angiogenesis by stimulating the production of thrombospondin. Cold Spring Harb Symp Quant Biol 59:483–489

Dameron KM, Volpert OV, Tainsky MA et al (1994) Control of angiogenesis in fibroblasts by p53 regulation of thrombospondin-1. Science 265:1582–1584

Darland DC, Massingham LJ, Smith RL et al (2003) Pericyte production of cell-associated VEGF is differentiation-dependent and is associated with endothelial survival. Dev Biol 264:275–288

Daugherty BL, Siciliano SJ, De Martino JA et al (1996) Cloning, expression, and characterization of the human eosinophil eotaxin receptor. J Exp Med 183:2349–2354

Davis S, Aldrich TH, Jones PF et al (1996) Isolation of angiopoietin-1, a ligand for the TIE2 receptor, by secretion-trap expression cloning. Cell 87:1161–1169

D'Amato RJ, Loughman MS, Flynn E et al (1994) Thalidomide is an inhibitor of angiogenesis. Proc Natl Acad Sci USA 91:4082–4085

De Palma M, Venneri NA, Galli R et al (2005) Tie-2 identifies a hematopoietic lineage of proangiogenic monocytes required for tumor vessel formation and a mesenchymal population of pericyte progenitors. Cancer Cell 8:211–226

de Vries C, Escobedo JA, Ueno H et al (1992) The fms-like tyrosine kinase, a receptor for vascular endothelial growth factor. Science 255:989–991

DeYoung B, Wick MR, Fitzgibbon JF et al (1993) CD31: an immunospecific marker for endothelial differentiation in human neoplasms. Appl Immunohistochem 1:97–100

Dell'Era P, Belleri M, Stabile H et al (2001) Paracrine and autocrine effects of fibroblast growth factor-4 in endothelial cells. Oncogene 20:2655–2663

Dell'Era P, Coco L, Ronca R et al (2002) Gene expression profile in fibroblast growth factor 2-transformed endothelial cells. Oncogene 21:2433–2440.

Dellacono F, Spiro J, Eisma R et al (1997) Expression of basic fibroblast growth factor and its receptors by head and neck squamous carcinoma tumor and vascular endothelial cells. Am J Surg 174:540–544

Denekamp J (1982) Endothelial cell proliferation as a novel approach to targeting tumor therapy. Br J Cancer 45:136–139

Denhardt DT, Noda M, O'Regan AW et al (2001) Osteopontin as a means to cope with environment insults: regulation of inflammation, tissue remodeling and cell survival. J Clin Invest 107:1055–1061

Dias S, Boyd R, Balkwill F (1998) IL-12 regulates VEGF and MMPs in a murine breast cancer model. Int J Cancer 78:361–365

Dietz A, Rudat V, Conradt C et al (2000) Prognostic relevance of serum levels of the angiogenic peptide bFGF in advanced carcinoma of the head and neck treated by primary radiochemotherapy. Head Neck 22:666–673

Di Carlo E, Forni G, Musiani P (2003) Neutrophils in the antitumoral immune response. Chem Immunol Allergy 83:182–203

Di Raimondo FD, Azzaro MP, Palumbo GA et al (2000) Angiogenic factors in multiple myeloma: higher levels in bone marrow that in peripheral blood. Haematologica 85:800–805

Dirix LY, Vermeulen PB, Pawinski A et al (1997) Elevated levels of the angiogenic cytokines basic fibroblast growth factor and vascular endothelial growth factor in sera of cancer patients. Br J Cancer 76:238–243

Dorrell MI, Aguilar E, Scheppke L et al (2007) Combination of angiostatic therapy complete inhibits ocular and tumor angiogenesis. Proc Natl Acad Sci USA 104:967–972

Dorta RG, Landman G, Kowalski LP et al (2002) Tumour-associated tissue eosinophilia as a prognostic factor in oral squamous cell carcinomas. Histopathology 41:152–157

Dredge K, Marriott JB, Macdonald CD et al (2002) Novel thalidomide analogous display anti-angiogenic activity independently of immunomodulatory effects. Br J Cancer 87:1166–1172

Duke-Elder S, Perkins ES (1996) Cysts and tumors of the uveal tract. In: Duke-Elder S (ed) System of ophthalmology. Mosby Company, St. Louis, CV, pp 754–937

Dumont DJ, Yamaguchi TP, Conlon RA et al (1992) Tek, a novel tyrosine kinase gene located on mouse chromosome 4, is expressed in endothelial cells, and their presumptive precursors. Oncogene 7:1471–1480

Dvorak HF, Dvorak AM, Simpson BA et al (1970) Cutaneous basophil hypersensitivity. II. A light and electron microscopic description. J Exp Med 132:558–582

Dvorak HF, Dvorak AM, Churchill WH (1973) Immunologic rejection of diethylnitrosamine-induced hepatomas in strain 2 guinea pigs: participation of basophilic leukocytes and macrophage aggregates. J Exp Med 137:751–755

Dvorak HF, Dvorak AM, Manseau EJ et al (1979a) Fibrin gel investment associated with line 1 and line 10 solid tumor growth, angiogenesis, and fibroplasia in guinea pigs. Role of cellular immunity, myofibroblasts, microvascular damage, and infarction in line 1 tumor regression. J Natl Cancer Inst 62:1459–1472

Dvorak HF, Orenstein NS, Carvalho AC et al (1979b) Induction of a fibrin-gel investment: an early event in line 10 hepatocarcinoma growth mediated by tumor-secreted products. J Immunol 122:166–174

Dvorak AM, Mihm MC Jr, Osage JE et al (1980) Melanoma. An ultrastructural study of the host inflammatory and vascular responses. J Invest Dermatol 75:388–393

Dvorak HF, Dickersin GR, Dvorak AM et al (1981) Human breast carcinoma: fibrin deposits and desmoplasia. Inflammatory cell type and distribution. Microvasculature and infarction. J Natl Cancer Inst 67:337–345

Dvorak HF, Senger DR, Dvorak AM (1983) Fibrin as a component of the tumor stroma: origins and biological significance. Cancer Metastasis Rev 2:41–73

Dvorak HF, Harvey VS, McDonagh J (1984) Quantitation of fibrinogen influx and fibrin deposition and turnover in line 1 and line 10 guinea pig carcinomas. Cancer Res 44:3348–3354

Dvorak HF (1986) Tumors: wounds that not heal. Similarities between tumor stroma generation and wound healing. N Engl J Med 315:1650–1659

Dvorak HF, Nagy JA, Dvorak JT et al (1988) Identification and characterization of the blood vessel of solid tumors that are leaky to circulating macromolecules. Am J Pathol 133:95–109

Dvorak HF, Gresser I (1989) Microvascular injury in pathogenesis of interferon-induced necrosis of subcutaneous tumors in mice. J Natl Cancer Inst 81:497–502

Dvorak HF, Nagy JA, Berse B et al (1992) Vascular permeability factor, fibrin, and the pathogenesis of tumor stroma formation. Ann NY Acad Sci 667:101–111

Eberhard A, Kahlert S, Goede V et al (2000) Heterogeneity of angiogenesis and blood vessel maturation in human tumors: implications for antiangiogenic tumor therapies. Cancer Res 60:1388–1393

Egami K, Murohara T, Shimada T et al (2003) Role of host angiotensin II type 1 receptor in tumor angiogenesis and growth. J Clin Invest 112:67–75

Ehrmann RL, Knoth M (1968) Choriocarcinoma: transfilter stimulation of vasoproliferation in the hamster cheek pouch studied by light and electron microscopy. J Natl Cancer Inst 41:1239–1241

Elpek GO, Gelen T, Aksoy NH et al (2001) The prognostic relevance of angiogenesis and mast cells in squamous cell carcinoma of the oesophagus. J Clin Pathol 54:940–944

Engsig MT, Chen QJ, Vu TH et al (2000) Matrix metalloproteinase-9 and vascular endothelial growth factor are essential for osteoclast recruitment into developing long bones. J Cell Biol 151:879–890.

Esch F, Baird A, Ling N et al (1985) Primary structure of bovine pituitary basic fibroblast growth factor (FGF) and comparison with the amino-terminal sequence of bovine brain acidic FGF. Proc Natl Acad Sci USA 81:6507–6511

Ezekowitz RA, Mulliken JB, Folkman J (1992) Interferon alfa-2a therapy for lifethreatening hemangiomas of infancy. N Engl J Med 326:1456–1463

Fan F, Wey JS, McCarty MF et al (2005) Expression and function of vascular endothelial growth factor receptor-1 on human colorectal cancer cells. Oncogene 24:2647–2653

Fenselau A, Kaiser D, Wallis K (1981) Nucleoside requirements for the *in vivo* growth of bovine aortic endothelial cells. J Cell Physiol 108:375–384

Fernandez-Aguilar S, Jondet M, Simonart T et al (2006) Microvessel and lymphatic density in tubular carcinoma of the breast: comparative study with invasive low-grade ductal carcinoma. Breast 15:782–785

Ferrara N, Henzel WJ (1989) Pituitary follicular cells secrete a novel heparin-binding growth factor specific for vascular endothelial cells. Biochem Biophys Res Commun 161:851–858

Ferrara N, Carver-Moore K, Chen H et al (1996) Heterozygous embryonic lethality induced by targeted inactivation of the VEGF gene. Nature 380:439–442

Ferrara N, Kerbel RS (2005) Angiogenesis as a therapeutic target. Nature 438:967–974

Flossmann E, Rothwell PM (2007) Effect of aspirin on long-term risk of colorectal cancer: consistent evidence from randomised and observational studies. Lancet 369:1603–1613

Folkman MJ, Long DM, Becker FF (1963) Growth and metastasis of tumor in organ culture. Cancer 16:453–467

Folkman J (1971) Tumor angiogenesis. Therapeutic implications. N Engl J Med 285:1182–1186

Folkman J, Merler E, Abernathy C et al (1971) Isolation of a tumor fraction responsible for angiogenesis. J Exp Med 133:275–288

Folkman J (1975) Tumor angiogenesis. In: Becker FF (ed) Cancer comprehensive treatise. Plenum Press, New York, pp 355–390

Folkman J, Haudenschild CC, Zetter BR (1979) Long-term culture of capillary endothelial cells. Proc Natl Acad Sci USA 76:5217–5221

Folkman J, Langer R, Linhardt R et al (1983) Angiogenesis inhibition and tumor regression caused by heparin or a heparin fragment in the presence of cortisone. Science 221:719–725

Folkman J, Hochberg M (1983) Self-regulation of growth in three dimensions. J Exp Med 138:745–753

Folkman J, Watson K, Ingber D et al (1989) Induction of angiogenesis during the transition from hyperplasia to neoplasia. Nature 339:58–61

Folkman J, Jalluri R (2003) Tumor angiogenesis. In: Kufe DW et al (ed) Cancer medicine, BC Decker, Hamilton Ontario, pp 161–194

Fox SB (1997) Tumour angiogenesis and prognosis. Histopathology 30:294–301

Francois J, Neetens A (1967) Physico-anatomical studies of spontaneous and experimental intraocular new growths: vascular supply. Bibl Anat 9:403–411

Fukushima N, Satoh T, Sano M et al (2001) Angiogenesis and mast cells in non-Hodgkin's lymphoma: a strong correlation in angioimmunoblastic T-cell lymphoma. Leuk Lymphoma 42:709–720

Gale NW, Thurston G, Hackett SF et al (2002) Angiopoietin-2 is required for postnatal angiogenesis and lymphatic patterning, and only the latter role is rescued by Angiopoietin-1. Dev Cell 3:411–423

Gao D, Nolan DJ, Mellick AS et al (2008) Endothelial progenitor cells control the angiogenic switch in mouse lung metastasis. Science 319:195–198

Garcia-Barros M, Paris F, Cordon-Cardo C et al (2003) Tumor response to radiotherapy regulated by endothelial cell apoptosis. Science 300:1155–1159

Gordan JD, Simon MC (2007) Hypoxia-inducible factors: central regulators of the tumor phenotype. Curr Opin Genet Dev 17:71–77

Gasparini G, Longo R, Fanelli M et al (2005) Combination of antiangiogenic therapy with other anticancer therapies: results, challenges, and open questions. J Clin Oncol 23:1295–1311

Gately S, Twardowski P, Tack MS et al (1997) The mechanism of cancer-mediated conversion of plasminogen to the angiogenesis inhibitor angiostatin. Proc Natl Acad Sci USA 94:10868–10872

Gehlin UM, Ergun S, Schumacher U et al (2000) *In vitro* differentiation of endothelial cells from AC 133 positive progenitor cells. Blood 95:3106–3112

Gerber HP, Condorelli F, Park J et al (1997) Differential transcriptional regulation of the two VEGF receptor genes. Flt-1, but not Flk-1/KDR is up-regulated by hypoxia. J Biol Chem 272:23659–23667

Gerber HP, Kowalski J, Sherman D et al (2000) Complete inhibition of rhabdomyosarcoma xenograft growth and neovascularization requires blockade of both tumor and host vascular endothelial growth factor. Cancer Res 60:6253–6258

Gerhardt H, Ruhrberg C, Abramsson A et al (2004) Neuropilin-1 is required for endothelial tip cell guidance in the developing central nervous system. Dev Dyn 231:503–509

Giancotti FG, Ruoslahti E (1999) Integrin signaling. Science 285:1028–1032

Gimbrone MA Jr, Aster RH, Cotran RS et al (1969) Preservation of vascular integrity in organs perfused in vitro with a platelet-rich medium. Nature 222:33–36

Gimbrone MA, Leapman SB, Cotran RS et al (1972) Tumor dormancy in vivo by prevention of neovascularization. J Exp Med 136:261–276

Gimbrone MA, Leapman S, Cotran RS et al (1973) Tumor angiogenesis: iris neovascularization at a distance from experimental intraocular tumors. J Natl Cancer Inst 50:219–228

Gimbrone MA Jr, Cotran RS, Folkman J (1974) Tumor growth and neovascularization: an experimental model using rabbit cornea. J Natl Cancer Inst 52:413–427

Gimbrone M, Gullino PM (1976a) Angiogenic capacity of preneoplastic lesions of the murine mammary gland as a marker of neoplastic transformation. Cancer Res 36:2611–2620

Gimbrone M, Gullino PM (1976b) Neovascularization induced by intraocular xenografts of normal, preneoplastic and neoplastic mouse mammary tissue. J Natl Cancer Inst 56:305–318

Ginns LC, Roberts DH, Mark EJ et al (2003) Pulmonary capillary hemangiomatosis with atypical endotheliomatosis: successful antiangiogenic therapy with doxycycline. Chest 124:2017–2022

Giraudo E, Inoue M, Hanahan D (2004) An amino-bisphosphonate targets MMP-9-expressing macrophages and angiogenesis to impair cervical carcinogenesis. J Clin Invest 114:623–633

Glade Bender J, Cooney EM, Kandek JJ et al (2004) Vascular remodeling and clinical resistance to antiangiogenic cancer therapy. Drug Resist Updat 7:289–300

Glowacki J, Mulliken JB (1982) Mast cells in haemangiomas and vascular malformations. Pediatrics 70:48–51

Goldman E (1907) The growth of malignant disease in man and the lower animals with special reference to the vascular system. Lancet ii:1236–1240

Good DJ, Polverini PJ, Rastinejad F et al (1990) A tumor suppressor-dependent inhibitor of angiogenesis is immunologically and functionally indistinguishable from a fragment of thrombospondin. Proc Natl Acad Sci USA 87:6624–6628

Gordon MS, Margolin K, Talpaz M et al (2001) Phase I safety and pharmacokinetic study of recombinant human anti-vascular endothelial growth factor in patients with advanced cancer. J Clin Oncol 19:843–850

Gospodarowicz D (1975) Purification of a fibroblast growth factor from bovine pituitary. J Biol Chem 250:2515–2520

Gospodarowicz D, Bialecki H, Greenburg G (1978) Purification of the fibroblast growth factor activity from bovine brain. J Biol Chem 253:3736–3743

Gospodarowicz D, Cheng J, Lui GM et al (1984) Isolation of brain fibroblast growth factor by heparin-sepharose affinity chromatography: identify with pituitary fibroblast growth factor. Proc Natl Acad Sci USA 81:6963–6967

Graeber TG, Osmanian C, Jacks T et al (1996) Hypoxia-mediated selection of cells with diminished apoptotic potential in solid tumours. Nature 379:88–91

Graham R, Graham J (1966) Mast cells and cancer of the cervix. Surg Gynecol Obstet 123:2–9

Grant MB, May WS, Caballero S et al (2002) Adult hematopoietic stem cells provide functional hemangioblast activity during retinal neovascularization. Nat Med 8:607–612

Green HSN (1941) Heterologous transplantation of mammalian tumors: I. The transfer of rabbit tumors to alien species. J Exp Med 73:461–474

Greenblatt M, Shubik P (1968) Tumor angiogenesis: transfilter diffusion studied in the hamster by the transparent chamber technique. J Natl Cancer Inst 41:111–124

Gross R, Dickson C (2005) Fibroblast growth factor signaling in tumorigenesis. Cytokine Growth Factor Rev 16:179–186

Gruber BL, Marchese MJ, Kaw R (1995) Angiogenic factors stimulate mast cell migration. Blood 86:2488–2493

Gruber M, Simon MC (2006) Hypoxia-inducible factors, hypoxia, and tumor angiogenesis. Curr Opin Hematol 13:169–174

Gualandris A, Rusnati M, Belleri M et al (1996) Basic fibroblast growth factor overexpression in endothelial cells: an autocrine mechanism for angiogenesis and angioproliferative diseases. Cell Growth Differ 7:147–160

Guidi AJ, Abu-Jawdeh G, Berse B et al (1995) Vascular permeability factor (vascular endothelial growth factor) expression and angiogenesis in cervical neoplasia. J Natl Cancer Inst 87:1237–1245

Guidi AJ, Berry DA, Broadwater G et al (2000) Association of angiogenesis in lymph node metastases with outcome of breast cancer. J Natl Cancer Inst 92:486–492

Gulati R, Jevremovic D, Peterson TE et al (2003) Diverse origin and function of cells with endothelial phenotype obtained from adult human blood. Circ Res 93:1023–1025

Guo P, Hu B, Gu W et al (1985) Platelet-derived growth factor-B enhances glioma angiogenesis by stimulating vascular endothelial growth factor expression in tumor endothelia and by promoting pericyte recruitment. Am J Pathol 162:1083–1093

Hammersen F, Endrich B, Messmer K (1985) The fine structure of tumor blood vessels, I. Participation of non-endothelial cells in tumor angiogenesis. Int J Microcirc Clin Exp 4:31–43

Hanada T, Nakagawa M, Emoto A et al (2000) Prognostic value of tumor-associated macrophage count in human bladder cancer. Int J Urol 7:263–269

Hanahan D (1985) Heritable formation of pancreatic β-cell tumors in transgenic mice expressing recombinant insulin/simian virus 40 oncogene. Nature 315:115–122

Hanahan D, Folkman J (1996) Patterns and emerging mechanisms of the angiogenic switch during tumorigenesis. Cell 86:353–364

Hanahan D (1997) Signaling vascular morphogenesis and maintenance. Science 277:48–50

Hanahan D, Bergers G, Bergsland E (2000) Less is more, regularly: metronomic dosing of cytotoxic drugs can target tumor angiogenesis in mice. J Clin Invest 105:1045–1047

Hanahan D, Weinberg RA (2000) The hallmarks of cancer. Cell 100:57–70

Hansen S, Grabaud DA, Rose C et al (1998) Angiogenesis in breast cancer. A comparative study of the observer variability of methods for determining microvessel density. Lab Invest 78:1563–1573

Hansen S, Grabaud DA, Sorensen FB et al (2000) The prognostic value of angiogenesis by Chalkley counting in a confirmatory study design on 836 breast cancer patients. Clin Cancer Res 6:139–146

Harris NL, Dvorak AM, Smith J et al (1982) Fibrin deposits in Hodgkin's disease. Am J Pathol 108:119–129

Hartveit F (1981) Mast cells and metachromasia in human breast cancer: their occurrence, significance and consequence: a preliminary report. J Pathol 134:7–11

Hattori K, Dias S, Heissig B et al (2001) Vascular endothelial growth factor and angiopoietin-1 stimulate postnatal hematopoiesis by recruitment of vasculogenic and hematopoietic stem cells. J Exp Med 193:1005–1014

He Y, kozaki K, Karpanen T et al (2002) Suppression of lymphangiogenesis and lymph node metastasis by blocking vascular endothelial growth factor receptor-3 signaling. J Natl Cancer Inst 94:819–925

He Y, Rajantie I, Ilmonen M et al (2004) Preexisting lymphatic endothelium but not endothelial progenitor cells are essential for tumor lymphangiogenesis and lymphatic metastasis. Cancer Res 64:737–3740

Heeschen C, Aicher A, Lehmann R et al (2003) Erythropoietin is a potent physiologic stimulus for endothelial progenitor cell mobilization. Blood 102:1340–1346

Hickey MM, Simon MC (2006) Regulation of angiogenesis by hypoxia and hypoxia-inducible factors. Curr Top Dev Biol 76:217–257

Hida K, Hida Y, Amin DN et al (2004) Tumor-associated endothelial cells with cytogenetic abnormalities. Cancer Res 64:8249–8255

Hida K, Hida Y, Shindoh M (2008) Understanding tumor endothelial cells abnormalities to develop ideal anti-angiogenic therapies. Cancer Sci 99:459–466

Hillen F, van de Winkel A, Creytens D et al (2006) Proliferating endothelial cells, but not microvessel density, are a prognostic parameter in human cutaneous melanoma. Melanoma Res 16:453–457

Hlatky L, Hahnfeldt P, Folkman J (2002) Clinical application of antiangiogenic therapy: microvessel density, what it does and doesn't tell us. J Natl Cancer Inst 94:883–893

Holash J, Wiegand SJ, Yancopoulos GF (1999) New model of tumor angiogenesis: dynamic balance between vessel regression and growth mediated by angiopoietins and VEGF. Oncogene 18:5356–5362

Holash J, Maisonpierre PC, Compton D et al (1999) Vessel cooption, regression and growth in tumors mediated by angiopoietins and VEGF. Science 284:1994–1998

Holash J, Davis S, Papadopoulos N et al (2002) VEGF-Trap: a VEGF blocker with potent antitumor effects. Proc Natl Acad Sci USA 99:11393–11398

Horiuchi T, Weller PF (1997) Expression of vascular endothelial growth factor by human eosinophils: upregulation by granulocyte macrophage colony stimulating factor and interelukin-5. Am J Respir Cell Mol Biol 17:70–77

Houck KA, Ferrara N, Winer J et al (1991) The vascular endothelial growth factor family: identification of a fourth molecular species and characterization of alternative splicing of RNA. Mol Endocrinol 5:1806–1814

Houck KA, Leung DW, Rowland AM et al (1992) Dual regulation of vascular endothelial growth factor bioavailability by genetic and proteolytic mechanisms. J Biol Chem 267:26031–26037

Hunter J (1794) Treatise on the blood, inflammation and gunshot wounds. Philadelphia, Thomas Bradford

Huntington GS, Mc Clure CFW (1910) The anatomy and development of the jugular lymph sac in the domestic cat (Felis domestica) Am J Anat 10:177–311

Hur J, Yoon CH, Kim HS et al (2004) Characterization of two types of endothelial progenitor cells and their different contributions to neovasculogenesis. Arterioscler Thromb Vasc Biol 24:288–293

Hurwitz H, Fehrenbacher L, Novotny W et al (2004) Bevacizumab plus irinotecan, fluorouracil, and leucovarin for metastatic colorectal cancer. New Engl J Med 350:2335–2342

Hurwitz H, Saini S (2006) Bevacizumab in the treatment of metastatic colorectal cancer: safety profile and management of adverse events. Semin Oncol 33:S26–S34

Ide AG, Baker NH, Warren SL (1939) Vascularization of the Brown-Pearce rabbit epithelioma transplant as seen in the transparent ear chambers. Am J Roentg 32:891–899

Imada D, Shijubo N, Koijma H et al (2000) Mast cells correlate with angiogenesis and poor outcome in stage I lung adenocarcinoma. Eur Resp J 15:1087–1093

Inai T, Mancuso M, Hashizume H et al (2004) Inhibition of vascular endothelial growth factor (VEGF) signaling in cancer causes loss of endothelial fenestrations, regression of tumor vessels and appearance of basement membrane ghosts. Am J Pathol 165:35–52

Ingber DE, Madri JA, Folkman J (1986) A possible mechanism for inhibition of angiogenesis by angiostatic steroids: induction of capillary basement membrane dissolution. Endocrinology 119:1768–1775

Ingber D, Fujita T, Kishimoto S et al (1990) Synthetic analogues of fumagillin that inhibit angiogenesis and suppress tumour growth. Nature 468:555–557

Inoue M, Hager JH, Ferrara N et al (2002) VEGF-A has a critical, nonredundant role in angiogenic switching and pancreatic beta cell carcinogenesis. Cancer Cell 1:193–202

Iruela-Arispe ML, Bornstein P, Sage H (1991) Thrombospondin exerts an antiangiogenic effect on cord formation by endothelial cells in vitro. Proc Natl Acad Sci USA 88:5026–5030

Isaka N, Padera TP, Hagendoorn J et al (2004) Peritumoral lymphatics induced by vascular endothelial growth factor-C exhibit abnormal function. Cancer Res 64:4400–4404

Ito H, Rovira II, Bloom ML et al (1999) Endothelial progenitor cells as putative targets for angiostatin. Cancer Res 59:5875–5877

Itoh N, Ornitz DM (2004) Evolution of the Fgf and Fgfr gene families. Trends Genet 20:563–569

Iwaguro H, Yamaguchi J, Kalka C et al (2002) Endothelial progenitor cell vascular endothelial growth factor gene transfer for vascular regeneration. Circulation 105:732–738

Jackson DG, Prevo R, Clasper S et al (2001) LYVE-1, the lymphatic system and tumor lymphangiogenesis. Trends Immunol 22:317–321

Jackson KA, Majka SM, Wang H et al (2001) Regeneration of ischemic cardiac muscle and vascular endothelium by adult stem cells. J Clin Invest 107:1395–1402

Jaffe EA, Nachman RL, Becker CG et al (1973) Culture of human endothelial cells derived from umbilical veins. Identification by morphologic and immunologic criteria. J Clin Invest 52:2745–2756

Jain RK (2005) Normalization of tumor vasculature: an emerging concept in antiangiogenic therapy. Science 307:58–62

Jain RK, Duda DG, Clark JW et al (2006) Lessons from phase III clinical trials on anti-VEGF therapy for cancer. Nat Clin Pract Oncol 3:24–40

Jeltsch M, Kaipainen A, Joukov V et al (1997) Hyperplasia of lymphatic vessels in VEGF-C transgenic mice. Science 276:1423–1425

Jenkins DC, Charles IG, Thompson LL et al (1995) Roles of nitric oxide in tumor growth. Proc Natl Acad Sci USA 92:4392–4396

Jensen OA (1976) The "Knapp-Ronne" type of malignant melanoma of the choroid: a haemangioma-like melanoma with a typical clinical picture. So-called "preretinal malignant choroid melanoma". Acta Ophthalmol 54:41–54

Johnson DE, Williams LT (1993) Structural and functional diversity in the FGF receptor multigene family. Adv Cancer Res 60:1–41

Joukov V, Pajusola K, Kaipainen A et al (1996) A novel vascular endothelial growth factor, VEGF-C, is a ligand for the Flt4 (VEGFR-3) and KDR (VEGFR-2) receptor tyrosine kinases. EMBO J 15: 1751

Joyce JA, Laakkonen P, Bernasconi M et al (2003) Stage-specific markers revealed by phage display in a mouse model of pancreatic islet tumorigenesis. Cancer Cell 4:393–403

Joyce JA, Baruch A, Chelade K et al (2004) Cathepsin cysteine proteases are effectors on invasive growth and angiogenesis during multistage tumorigenesis. Cancer Cell 5:443–453

Joyce JA, Freeman C, Meyer-Morse N et al (2005) A functional heparan sulfate mimetic implicated both heparanase and heparan sulfate in tumor angiogenesis and invasion in a mouse model of multistage cancer. Oncogene 24:4037–4051

Jussila L, Alitalo K (2002) Vascular growth factors and lymph angiogenesis. Physiol Rev 8:673–700

Kabbinavar F, Hurwitz HI, Fehrenbacher L et al (2003) Phase II, randomized trial comparing bevacizumab plus fluorouracil (FU)/leucovorin (LV) with FU/LV alone in patients with metastatic colorectal cancer. J Clin Oncol 21:60–65

Kaban LB, Mulliken JB, Ezekowiyz RA et al (1999) Antiangiogenic therapy of a recurrent giant cell tumor of the mandible with interferon alpha-2a. Pediatrics 103:1145–1149

Kaban LB, Troulis MJ, Ebb D et al (2002) Antiangiogenic therapy with interferon alpha for giant cell lesions of the jaws. J Oral Maxillofac Surg 60:1103–1111

Kalka C, Masuda H, Takahashi T et al (2000) Transplantation of ex vivo expanded endothelial progenitor cells for therapeutic neovascularization. Proc Nat Acad Sci USA 97:3422–3427

Kalluri R (2003) Basement membranes: structure, assembly and role in tumour angiogenesis. Nat Rev Cancer 3:422–433

Kamen BA, Glod J, Cole PD (2006) Metronomic therapy from a pharmacologist's view. J Pediatr Hematol Oncol 28:325–327

Kandel J, Bossy-Wetzel E, Radvanyi F et al (1991) Neovascularization is associated with a switch to the export of bFGF in the multistep development of fibrosarcoma. Cell 66:1095–1104

Karkkainen MJ, Saaristo A, Jussila L et al (2001) A model for gene therapy of human hereditary lymphedema. Proc Natl Acad Sci USA 98:12677–12682

Karkkainen MJ, Haiko P, Sainio K et al (2004) Vascular endothelial growth factor C is required for sprouting of the first lymphatic vessels from embryonic veins. Nat Immunol 5:74–80

Kerbel RS, Yu J, Tran J et al (2001) Possible mechanisms of acquired resistance to anti-angiogenic drugs: implications for the use of combination therapy approaches. Cancer Metastasis Rev 20:79–86

Kessler DA, Langer RS, Pless NA, Folkman J (1976) Mast cells and tumor angiogenesis. Int J Cancer18:703–709

Kim KJ, Li B, Winer J et al (1993) Inhibition of vascular endothelial growth factor-induced angiogenesis suppresses tumor growth in vivo. Nature 362:841–844

Kirsch M, Schackert G, Black PM (2004) Metastasis and angiogenesis. Cancer Treat Res 117:285–304

Kisker O, Becker CM, Prox D et al (2001) Continuous administration of endostatin by intraperitoneally implanted osmotic pump improves the efficacy and potency of therapy in a mouse xenograft tumor model. Cancer Res 61:7669–7774

Klein B, Zhang X, Lu Z (1995) Interleukin-6 in human multiple myeloma. Blood 85:863–872

Klement G, Baruchel S, Rak J et al (2000) Continuous low-dose therapy with vinblastine and VEGF receptor-2 antibody induces sustained tumor regression without overt toxicity. J Clin Invest 105:R15–R24

Klimp AH, Hollema H, Kempinga C et al (2001) Expression of cyclooxygenase-2 and inducible nitric oxide synthase in human ovarian tumors and tumor-associated macrophages. Cancer Res 61:7305–7309

Knighton D, Ausprunk D, Tapper D et al (1977) Avascular and vascular phases of tumour growth in the chick embryo. Br J Cancer 35:347–356

Kochne CH, Dubois RN (2004) COX-2 inhibition and colorectal cancer. Semin Oncol 31:12–21

Koide N, Nishio A, Sato T et al (2004) Significance of macrophage chemoattractant protein-1 expression and macrophage infiltration in squamous cell carcinoma of the esophagus. Am J Gastroenterol 99:1667–1674

Korkolopoulou P, Apostolidou E, Pavlopoulos PM et al (2001) Prognostic evaluation of the microvascular network in myelodysplastic syndromes. Leukemia 15:1369–1376

Korkolopoulou P, Thymara I, Kavantzas N et al (2005) Angiogenesis in Hodgkin's lymphoma: a morphometric approach in 286 patients with prognostic implications. Leukemia 19:894–900

Koster A, Reamaekers JMM (2005) Angiogenesis in malignant lymphoma. Curr Opin Oncol 17:611–616

Kukk E, Lymboussaki A, Taira S et al (1996) VEGF-C receptor binding and pattern of expression with VEGFR-3 suggests a role in lymphatic vascular development. Development 122:3829–3837

Kumar S, Ghellal A, Li C et al (1999) Breast carcinoma: vascular density determined using CD105 antibody correlates with tumor prognosis. Cancer Res 59:856–861

Kusamura S, Derchain S, Alvarenga M et al (2003) Expression of p53, c-erb-2, Ki-67, and CD34 in granulosa cell tumor of the ovary. Int J Gynecol Cancer 13:450–457

Laakkonen P, Waltari P, Holopainen T et al (2007) Vascular endothelial growth factor receptor 3 is involved in tumor angiogenesis and growth. Cancer Res 67:593–599

Lachter J, Stein M, Lichting C et al (1995) Mast cells in colorectal neoplasis and premalignant disorders. Dis Colon Rectum 38:290–293

Lackner C, Jukic Z, Tsybrovskyy O et al (2004) Prognostic relevance of tumour-associated macrophages and von Willebrand factor-positive microvessels in colorectal cancer. Virchows Arch 445:160–167

Langer R, Folkman J (1976) Polymers for the sustained release of proteins and other macromolecules. Nature 263:797–800

Langer R, Cohn M, Vacanti J et al (1980) Control of tumor growth in animals by infusion of an angiogenesis inhibitor. Proc Natl Acad Sci USA 77:4331–4335

Laquente B, Vinals F, Germa JR (2007) Metronomic chemotherapy: an antiangiogenic scheduling. Clin Transl Oncol 9:93–98

Lawler J (2002) Thrombospondin-1 as an endogenous inhibitor of angiogenesis and tumor growth. J Cell Mol Med 6:1–12

Leali D, Dell'Era P, Stabile H et al (2003) Osteopontin (Eta-1) and fibroblast growth factor-2 crosstalk in angiogenesis. J Immunol 171:1085–1093

Leek RD, Lewis CE, Whitehouse R et al (1995) Association of macrophage infiltration with angiogenesis and progression in invasive breast carcinoma. Cancer Res 56:4625–4629

Leek RD, Landers RJ, Harris AL et al (1999) Necrosis correlate with high vascular density and focal macrophage infiltration in invasive carcinoma of the breast. Br J Cancer 79:991–995

Lee LF, Hellendall RP, Wang Y et al (2000) IL-8 reduced tumorigenicity of human ovarian cancer *in vivo* due to neutrophil infiltration. J Immunol 164:2769–2775

Leeven P, Perny M, Gebre-Medhin S et al (1994) Mice deficient for PDGF-B show renal, cardiovascular and hematological abnormalities. Genes Dev 8:1875–1887

Leibovich SJ, Polverini PJ, Shepard HM et al (1987) Macrophage-induced angiogenesis is mediated by tumour necrosis factor-alpha. Nature 329:630–632

Lengauer C, Kinzler KW, Vogelstein B (1998) Genetic instabilities in human cancers. Nature 396:643–649

Lenz HJ (2005) Antiangiogenic agents in cancer therapy Oncology 19:17–25

Letilovic T, Vrhovac R, Verstovsek S et al (2006) Role of angiogenesis in chronic lymphocytic leukemia. Cancer 107:925–934

Leung DW, Cachianes G, Kuang WJ et al (1989) Vascular endothelial growth factor is a secreted angiogenic mitogen. Science 246:1306–1309

Lewis CE, Leek R, Harris A et al (1995) Cytokine regulation of angiogenesis in breast cancer: the role of tumor-associated macrophages. J Leukoc Biol 57:747–751

Lewis CE, Murdoch C (2005) Macrophages responses to hypoxia: implications for tumor progression and anticancer therapies. Am J Pathol 167:627–635

Li VW, Folker RD, Watanabe H et al (1994) Microvessel count and cerebrospinal fluid basic fibroblast growth factor in children with brain tumours. Lancet 344:82–86

Lin EY, Nguyen AV, Russel RG et al (2001) Colony-stimulating factor 1 promotes progression of mammary tumors to malignancy. J Exp Med 193:727–740

Lin EY, Lin JF, Gnatovski YL et al (2006) Macrophages regulate the angiogenic switch in a mouse model of human breast cancer. Cancer Res 66:11238–11246

Lin Y, Weisdorf DJ, Solovey A et al (2000) Origins of circulating endothelial cells and endothelial outgrowth from blood. J Clin Invest 105:71–77

Lindahl P, Johansson BR, Leveen P et al (1997) Pericyte loss microaneurysm formation in PDGF-B deficient mice. Science 277:242–245

Lindmark G, Gerdin B, Sunberg C et al (1996) Prognostic significance of the microvascular count in colorectal cancer. J Clin Oncol 14:461–466

Lissbrant IF, Stattin P, Wikstrom P et al (2000) Tumor associated macrophages in human prostate cancer: relation to clinicopathological variables and survival. Int J Oncol 17:445–451

Loges S, Heil G, Bruweleit M et al (2005) Analysis of concerted expression of angiogenic growth factors in acute myeloid leukemia: expression of angiopoietin-2 represents an independent prognostic factor for overall survival. J Clin Oncol 23:1109–1117

Looi LM (1987) Tumor-associated eosinophilia in nasopharingeal carcinoma. A pathologic study of 422 primary and 138 metastatic tumors. Cancer 59: 466–470

Lundberg LG, Hellstrom-Lindberg E, Kanter-Lewensohn L et al (2006) Angiogenesis in relation to clinical stage, apoptosis and prognostic score in myelodysplastic syndromes. Leuk Res 30:247–253

Lyden D, Hattori K, Dias S et al (2001) Impaired recruitment of bone marrow derived endothelial and hematopoietic precursor cells blocks tumor angiogenesis and growth. Nat Med 7:1194–1201

Lewis WH (1927) The vascular patterns of tumors. John Hopkins Hosp Bull 41:156–162

Li VW, Folkerth RD, Watanabe H et al (1994) Microvessel count and cerebrospinal fluid basic fibroblast growth factor in children with brain tumours. Lancet 344:82–86

Lopez T, Hanahan D (2002) Elevated levels of IGF-1 receptor convey invasive and metastatic capability in a mouse model of pancreatic islet tumorigenesis. Cancer Cell 1:339–353

Mabjeesh NJ, Amir S (2007) Hypoxia-inducible factor (HIF) in human tumorigenesis. Histol Histopathol 22:559–572

Machein MR, Knedla A, Knoth R et al (2004) Angiopoietin-1 promotes tumor angiogenesis in a rat glioma model. Am J Pathol 165:1557–1570

Magari S, Asano S (1978) Regeneration of the deep cervical lymphatics. Light and electron microscopic observations. Lymphology 11:57–61

Magari S (1987) Comparison of fine structure of lymphatics and blood vessels in normal conditions and during embryonic development and regeneration. Lymphology 20:189–195

Magnussen A, Kasman IM, Norberg S et al (2005) Rapid access of antibodies to $\alpha5\beta1$ integrin overexpressed on the luminal surface of tumor blood vessels. Cancer Res 65:2712–2721

Maione TE, Gray GS, Petro J et al (1990) Inhibition of angiogenesis by recombinant human platelet factor-4 and related peptides. Science 247:77–79

Maiorana A, Gullino PM (1978) Acquisition of angiogenic capacity and neoplastic transformation in the rat mammary gland. Cancer Res 38:4409–4414

Maisonpierre PC, Suri C, Jones PF et al (1997) Angiopoietin-2, a natural antagonist for Tie-2 that disrupts *in vivo* angiogenesis. Science 277:55–60

Makinen T, Veikkola T, Mustjoki S et al (2001) Isolated lymphatic endothelial cells transduce growth, survival and migratory signals via the VEGFC/D receptor VEGFR-3. EMBO J 20:4762–4773

Makitie T, Summanen P, Takkanen A et al (2001) Tumor-infiltrating macrophages (AD 68(+) cells) and prognosis in malignant uveal melanoma. Invest Ophtalmol Vis Sci 42:1414–1421

Mancuso P, Burlini A, Pruneri G et al (2001) Resting and activated endothelial cells are increased in the peripheral blood of cancer patients. Blood 97:3658–3661

Mandriota SJ, Jussila L, Jeltsch M et al (2001) Vascular endothelial growth factor-C-mediated lymphangiogenesis promotes tumour growth. EMBO J 20:672–682

Maniotis AJ, Folberg R, Hess A et al (1999) Vascular channel formation by human melanoma cells in vivo and in vitro: vasculogenic mimicry. Am J Pathol 155:739–752

Mantovani A, Bottazzi B, Colotta F et al (1992) The origin and function of tumor-associated macrophages. Immunol Today 13:265–270

Mantovani A, Sozzani S, Locati M et al (2002) Macrophage polarization: tumor-associated macrophages as a paradigm for polarized M2 mononuclear phagocytes. Trends Immunol 23:549–555

Margolin K, Gordon MS, Holmgren E et al (2001) Phase Ib trial of intravenous recombinant humanized monoclonal antibody to vascular endothelial growth factor in combination with chemotherapy in patients with advanced cancer: pharmacologic and long-term safety data. J Clin Oncol 19:851–856

Marioni G, Gaio E, Giacomelli L et al (2005) Endoglin (CD105) expression in head and neck basaloid squamous cell carcinoma. Acta Otolaryngol 125:307–311

Marioni G, Marino F, Giacomelli L et al (2006) Endoglin expression is associated with poor oncologic outcome in oral and oropharyngeal carcinoma. Acta Otolaryngol 126:633–639

Marrogi AJ, Travis WD, Welsh Ja et al (2000) Nitric oxide synthase, cyclooxygenase 2, and vascular endothelial growth factor in the angiogenesis of non-small cell lung carcinoma. Clin Cancer Res 6:4739–4744

Maula SM, Luukaa M, Grénman R et al (2003) Intratumoral lymphatics are essential for the metastatic spread and prognosis in squamous cell carcinomas of the head and neck region. Cancer Res 63:1920–1926

Mc Auslan BR, Hoffman H (1979) Endothelium stimulating factor from Walker carcinoma cells. Relation to tumor angiogenic factor. Exp Cell Res 119:181–190

Mc Donald DM, Munn L, Jain RK (2000) Vasculogenic mimicry: how convincing, how novel, and how significant. Am J Pathol 156:383–388

Mc Donald DM, Choyke PL (2003) Imaging of angiogenesis: from microscope to clinic. Nat Med 9:713–725

Mellin GW, Katzenstein M (1962) The saga of thalidomide. N Engl J Med 267:1184–1193

Merwin RM, Algire GH (1956) The role of graft and host vessels in vascularization of grafts of normal and neoplastic tissues. J Natl Cancer Inst 17:23–33

Mesiano S, Ferrara N, Jaffe RB (1998) Role of vascular endothelial growth factor in ovarian cancer: inhibition of ascite formation by immunoneutralization. Am J Pathol 153:1249–1256

Meyer M, Clauss M, Lepple-Wienhues A et al (1999) A novel vascular endothelial growth factor encoded by Orf virsu, VEGF-E, mediated angiogenesis via signalling through VEGFR-2 (KDR) but not VEGFR-1 (Flt-1) receptor tyrosine kinases. EMBO J 18:363–374

Michaelson LC (1948) The mode of development of the vascular system of the retina with some observations on its significance for certain retinal disorders. Trans Ophthalmol Soc UK, 68:137–180

Mikhail MS, Palan PR, Basu J et al (2004) Computerized measurement of intercapillary distance using image analysis in women with cervical intraepithelial neoplasia: correlation with severity. Acta Obstet Gynecol Scand 83:308–310

Miles AA, Miles EM (1952) Vascular reaction to histamine, histamine-liberator, and leukotaxine in the skin of guinea pigs. J Physiol 118:228–257

Milkowski DM, Weiss RA (1999) TNP 470. In: Teicher BA (ed) Antiangiogenic Agents in Cancer Therapy, Humana Press, Totowa, NJ, pp 385–398

Mimura K, Kono K, Takahashi A et al (2007) Vascular endothelial growth factor inhibits the function of human mature dendritic cells mediated by VEGF receptor-2. Cancer Immunol Immunother 56:761–770

Minhajat R, Mori D, Yamasaki F et al (2006) Endoglin (CD105) expression in angiogenesis of colon cancer: analysis using tissue microarrays and comparison with other endothelial markers. Virchows Arch 448:127–134

Miraglia S, Godfrey W, Yin AH et al (1997) A novel five-transmembrane hematopoietic stem cell antigen: isolation, characterization and molecular cloning. Blood 90:5013–5021

Mittal V, Nolan DJ (2007) Genomics and proteomics approaches in understanding tumor angiogenesis. Exper Rev Mol Diagn 7:133–147

Mizoguchi H, O'Shea JJ, Longo DL et al (1992) Alterations in signal transduction molecules in T lymphocytes from tumor-bearing mice. Science 258:1795–1798

Modzelewski RA, Davies P, Watkins SC et al (1994) Isolation and identification of fresh tumor-derived endothelial cells from a murine RIF-1 fibrosarcoma. Cancer Res 54:336–339

Mohle R, Bautz F, Rafii S et al (1998) The chemokine receptor CXCR-4 is expressed on CD34+ hematopoietic progenitors and leukemic cells and mediates transendothelial migration induced by stromal cell-derived factor-1. Blood 91:4523–4530

Molica S, Vitelli C, Levato D, et al (1999) Increased serum levels of vascular endothelial growth factor predict risk of progression in early B-cell chronic lymphocytic leukemia. Brit J Haematol 107:605–610

Molica S, Vacca A, Ribatti D et al (2002) Prognostic value of enhanced bone marrow angiogenesis in early B-cell chronic lymphocytic leukemia. Blood 100:3344–3351

Molica S, Vacca A, Crivellato E et al (2003) Tryptase-positive mast cells predict clinical outcome of patients with early B-cell chronic lymphocytic leukaemia. Eur J Haematol 71:137–139

Monestiroli S, Mancuso P, Burlini A et al (2001) Kinetics and viability of circulating endothelial cells as surrogate angiogenesis marker in an animal model of human lymphoma. Cancer Res 61:4341–4344

Morikawa S, Baluk P, Kaido H et al (2002) Abnormalities in pericytes on blood vessels and endothelial sprouts in tumors. Am J Pathol 160:985–1000

Moroni E, Dell'Era P, Rusnati M et al (2002) Fibroblast growth factors and their receptors in hematopoiesis and hematological tumors. J Hematother Stem Cell Res 11:19–32

Moses MA, Sudhalter J, Langer R (1990) Identification of an inhibitor of neovascularization from cartilage. Science 248:1408–1410

Moscatelli D, Presta M, Rifkin DB (1986a) Purification of a factor from human placenta that stimulates capillary endothelial cell protease production, DNA synthesis, and migration. Proc Natl Acad Sci USA 83:2091–2095

Moscatelli D, Presta M, Joseph-Silverstein J et al (1986b) R. Both normal and tumor cells produce basic fibroblast growth factor. J Cell Physiol 129:273–276

Mouta-Carreira C, Nasser SM, di Tomaso E et al (2001) LYVE-1 is not restricted to the lymph vessels: expression in normal liver blood sinusoids and down-regulation in human liver cancer cirrhosis. Cancer Res 61:8079–8084

Muhs BE, Plitas G, Delgado Y et al (2003) Temporal expression and activation of matrix metalloproteinases-2, -9 and membrane type 1-matrix metalloproteinase following acute hindlimb ischemia. J Surg Res 111:8–15

Nagy JA, Masse EM, Herzberg KT et al (1995) Pathogenesis of ascites tumor growth. Vascular permeability factor, vascular hyperpermeability and ascites fluid accumulation. Cancer Res 55:360–368

Nagy A, Vasile E, Feng D et al (2002) Vascular permeability factor/vascular endothelial growth factor induces lymphangiogenesis as well as angiogenesis. J Exp Med 196:1497–1506

Nakao S, Kuwano T, Tsutsumi-Miyahara C et al (2005) Infiltration with COX-2-expressing macrophages is a prerequisite for IL-1 beta-induced neovascularization and tumor growth. J Clin Invest 115:2979–2991

Naldini A, Carraro F (2005) Role of inflammatory mediators in angiogenesis. Curr Drug Targets Inflamm Allergy 4:3–8

Naldini A, Leali D, Pucci A et al (2006) Cutting edge: IL-1 beta mediates the proangiogenic activity of osteopontin-activated human monocytes. J Immunol 177:4267–4270

Naresh NK, Nerurkar AY, Borges AM (2001) Angiogenesis is a redundant for tumour growth in lymph node metastases. Histopathology 38:466–470

Negus RPK, Stamp GWQ, Hadlej J et al (1997) Quantitative assessment of the leucocyte infiltrate in ovarian cancer and its relationship to the expression of C-C chemokines. Am J Pathol 150:1723–1734

Nesbit M, Schaider H, Miller TH et al (2001) Low level monocyte chemoattractant protein-1 stimulation of monocytes leads to tumor formation in nontumorigenic melanoma cells. J Immunol 166:6483–6490

Nguyen M, Watanabe H, Budson AE et al (1994) Elevated levels of an angiogenic peptide, basic fibroblast growth factor, in the urine of patients with a wide spectrum of cancers. J Natl Cancer Inst 86:356–361

Nickoloff B, Ben-Neriah Y, Pikarsky E (2005) Inflammation and cancer: is the link as simple as we think? J Invest Dermatol 124:X–XIV

Nico B, Mangieri D, Crivellato E et al (2008a) Mast cells contribute to vasculogenic mimicry in multiple myeloma. Stem Cells Dev 17:850–854

Nico B, Mangieri D, Benagiamo V et al (2008b) Nerve growth factor as an angiogenic factor. Microvasc Res 75:135–141

Nicosia RF, Tchao R, Leighton J (1986) Interactions between newly formed endothelial channels and carcinoma cells in plasma clot cultures. Clin Exp Metastasis 4:91–104

Nielsen BS, Sehested M, Kjeldsen L et al (1997) Expression of matrix metalloprotease-9 in vascular pericytes in human breast cancer. Lab Invest 77:345–355

Nielsen JH, Hansen U, Christensen IJ et al (1999) Independent prognostic value of eosinophil and mast cell infiltration in colorectal cancer tissue. J Pathol 189:487–495

Nishie A, Ono M, Shono T et al (1999) Macrophage infiltration and heme oxygenase-1 expression correlate angiogenesis in human gliomas. Clin Cancer Res 5:1107–1113

Nozawa H, Chiu C, Hanahan D (2006) Infiltrating neutrophils mediate the initial angiogenic switch in a mouse model of multistage carcinogenesis. Proc Natl Acad Sci USA 103:12493–12498

Nowak G, Karrar A, Holmen C, et al (2004) Expression of vascular endothelial growth factor receptor-2 or Tie-2 on peripheral blood cells defines functionally competent cell population capable of reendothelialization. Circulation 110:3699–3707

Ny A, Koch M, Schneider M et al (2005) A genetic *Xenopus laevis* tadpole mode to study lymphangiogenesis. Nat Med 11:998–1004

O'Reilly MS, Holmgren L, Shing Y et al (1994) Angiostatin: a novel angiogenesis inhibitor that mediates the suppression of metastases by a Lewis lung carcinoma. Cell 79:315–328

O'Reilly MS, Holmgren L, Chen C et al (1996) Angiostatin induces and sustains dormancy of human primary tumors in mice. Nat Med 2:689–692

O'Reilly MS, Boehm T, Shing Y et al (1997) Endostatin: an endogenous inhibitor of angiogenesis and tumor growth. Cell 88:277–285

Odén B (1960) A microlymphoangiographic study of experimental wounds healing by second intention. Acta Chir Scand 120:100–114

Ohno S, Ohno Y, Suzuki N et al (2004) Correlation of histological localization of tumor-associated macrophages with clinicopathological features in endometrial cancer. Anticancer Res 24:3335–3342

Ohta T, Yamamoto M, Numata M et al (1995) Expression of basic fibroblast growth factor and its receptor in human pancreatic carcinomas. Br J Cancer 72:824–831

Orlandini M, Marconcini L, Ferruzzi R et al (1996) Identification of a *c-fos*-induced gene that is related to the platelet-derived growth factor/vascular endothelial growth factor family. Proc Natl Acad Sci USA 93:11675–11680

Orlidge A, D'Amore PA (1987) Inhibition of capillary endothelial cell growth by pericytes and smooth muscle cells. J Cell Biol 105:1455–1462

Park JE, Chen HH, Winer J et al (1994) Placenta growth factor. Potentiation of vascular endothelial growth factor bioactivity, in vitro and in vivo, and high affinity binding to Flt-1 but not to Flk-1/KDR. J Biol Chem 269:25646–25654

Partanen J, Armstrong E, Makela TP et al (1992) A novel endothelial cell surface receptor tyrosine kinase with extracellular epidermal growth factor homology domains. Mol Cell Biol 12:1698–1707

Pasqualini R, Arap W (2002) Translation of vascular diversity into targeted therapeutics. Ann Hematol 81(Suppl. 2):S66–S67

Pastorino U, Andreola S, Tagliabue E et al (1997) Immunohistochemical markers in stage I lung cancer: relevance to prognosis. J Clin Oncol 15:2858–2865

Peichev M, Naiyer AJ, Pereira D et al (2000) Expression of VEGFR-2 and AC133 by circulating human CD34(+) cells identifies a population of functional endothelial precursors. Blood 95:952–958

Pepper MS (2001) Lymphangiogenesis and tumor metastasis: myth or reality? Clin Cancer Res 7:462–468

Perez-Atayde AR, Sallan SE, Tedrow U et al (1997) Spectrum of tumor angiogenesis in the bone marrow of children with acute lymphocytic leukemia. Am J Pathol 150:815–821

Peters K, De Vries C, Williams L (1993) Vascular endothelial growth factor receptor expression during embryogenesis and tissue repair suggests a role in endothelial differentiation and blood vessel growth. Proc Natl Acad Sci USA 90:8915–8919

Pezzella F, Pastorino U, Tagliabue E et al (1997) Non-small-lung carcinoma tumor growth without morphological evidence of neo-angiogenesis. Am J Pathol 151:1417–1423

Pezzella F (2000) Evidence for novel non-angiogenic pathway in breast-cancer metastasis. Breast Cancer Progression Working Party. Lancet 355:1787–1788

Poole TJ, Zetter BR (1983) Stimulation of rat peritoneal mast cell migration by tumor-derived peptides. Cancer Res 43:5857–5861

Prause JU, Jensen OA (1980) Scanning electron microscopy of frozen-cracked, dry-cracked and enzyme-digested tissue of human malignant choriodal melanomas. Albrecht Von Gralfes Arch Klin Exp Ophthalmol 212:217–225

Presta LG, Chen H, O'Connor SJ et al (1997) Humanization of anti-vascular endothelial growth factor monoclonal antibody for the therapy of solid tumors and other disorders. Cancer Res 57:4593–4599

Presta M, Moscatelli D, Joseph-Silverstein J et al (1986) Purification from a human hepatoma cell line of a basic fibroblast growth factor-like molecule that stimulates capillary endothelial cell plasminogen activator production, DNA synthesis, and migration. Mol Cell Biol 6:4060–4066

Presta M, Dell'Era P, Mitola S et al (2005) Fibroblast growth factor/Fibroblast growth factor receptor system in angiogenesis. Cytokine Growth Factor Rev 16:159–178

Purhonen S, Palm J, Rossi D et al (2008) Bone marrow-derived circulating precursors do not contribute to vascular endothelium and are not needed for tumor growth. Proc Natl Acad Sci USA 105:6620–6625

Puxeddu I, Alian A, Piliponsky Am et al (2005a) Human peripheral blood eosinophils induce angiogenesis. Int J Biochem Cell Biol 37:628–636

Puxeddu I, Ribatti D, Crivellato E et al (2005b) Mast cells and eosinophils: a novel link between inflammation and angiogenesis in allergic diseases. J Allergy Clin Immunol 116:531–536

Queen MM, Ryan RE, Holzer RG et al (2005) Breast cancer cells stimulate neutrophils to produce oncostatin M: potential implications for tumor progression. Cancer Res 65:8869–8904

Rabtish W, Sperr WR, Lechner K et al (2004) Bone marrow microvessel density and its prognostic significance in AML. Leuk Lymphoma 45:1369–1373

Rafii S, Lyden D, Benezra R et al (2002) Vascular and haematopoietic stem cells: novel targets for antiangiogenesis therapy? Nat Rev Cancer 2:826–835

Rak J, Yu JL, Kerbel RS et al (2002) What do oncogenic mutations have to do with angiogenesis/vascular dependence of tumors? Cancer Res 62:1931–1934

Rastinejad F, Polverini PJ, Bouck NP (1989) Regulation of the activity of a new inhibitor of angiogenesis by a cancer suppressor gene. Cell 56:345–355

Reed JA, McNutt NS, Bogdany JK et al (1996) Expression of the mast cell growth factor interleukin-3 in melanocytic lesions correlates with an increased number of mast cells in the perilesional stroma: implications for melanoma progression. J Cutan Pathol 23:495–505

Rehman J, Li J, Orschell CM et al (2003) Peripheral blood "endothelial progenitor cells" are derived from monocyte/macrophages and secrete angiogenic growth factors. Circulation 107:1164–1169

Reinmuth N, Liu W, Jung YD et al (2001) Induction of VEGF in perivascular cells defines a potential paracrine mechanism for endothelial cell survival. FASEB J 15:1239–1241

Relf M, Le Jeune S, Scott PA et al (1997) Expression of the angiogenic factors vascular endothelial growth factor, acidic and basic fibroblast growth factor, transforming growth factor β-1, platelet-derived endothelial cell growth factor, placenta growth factor, and pleiotrophin in human primary breast cancer and its relation to angiogenesis. Cancer Res 57:963–969

Reyes M, Lund T, Lenvik T et al (2001) Purification and ex vivo expansion of postnatal human marrow mesodermal progenitor cells. Blood 98:2615–2625

Reyes M, Dudek A, Jahagirdar B et al (2002) Origin of endothelial progenitors in human postnatal bone marrow. J Clin Invest 109:337–346

Rhee J, Hoff PM (2005) Angiogenesis inhibitors in the treatment of cancer. Expert Opin Pharmacother 6:1701–1711

Rhodin JAG, Fujita H (1989) Capillary growth in mesentery of normal young rats. Intravital video and electron microscope analyses. J Submicrosc Cytol Pathol 21:1–34

Ribatti D, Vacca A, Nico B et al (1996) Angiogenesis spectrum in the stroma of B-cell non-Hodgkin's lymphoma. An immunohistochemical and ultrastructural study. Eur J Haematol 56:45–53

Ribatti D, Nico B, Vacca A et al (1998) Do mast cells help to induce angiogenesis in B cell non-Hodgkin's lymphomas? Br J Cancer 77:1900–1906

Ribatti D, Vacca A, Nico B et al (1999a) Bone marrow angiogenesis and mast cell density increase simultaneously with progression of human multiple myeloma. Br J Cancer 79:451–455

Ribatti D, Gualandris A, Belleri M et al (1999b) Alterations of blood vessel development by endothelial cells overexpressing fibroblast growth factor-2. J Pathol 189:590–599

Ribatti D, Vacca A, Marzullo A et al (2000) Angiogenesis and mast cell density with tryptase activity increase simultaneously with pathological progression in B-cell non-Hodgkin's lymphomas. Int J Cancer 82:171–175

Ribatti D, Polimeno G, Vacca A et al (2002) Correlation of bone marrow angiogenesis and mast cells with tryptase activity in myelodysplastic syndromes. Leukemia 16:1680–1684

Ribatti D, Molica S, Vacca A et al (2003 a) Tryptase-positive mast cells correlate positively with bone marrow angiogenesis in B-cell chronic lymphocytic leukaemia. Leukemia 17:1428–1430

Ribatti D, Vacca A, Ria R et al (2003 b) Neovascularization, expression of fibroblast growth factor-2, and mast cells with tryptase activity increase simultaneously with pathological progression in human malignant melanoma. Eur J Cancer 39:666–675

Ribatti D, Ennas MG, Vacca A et al (2003 c) Tumor vascularity and tryptase-positive mast cells correlate with a poor prognosis in melanoma. Eur J Clin Invest 33:420–425

Ribatti D, Vacca A, Dammacco F (2003 d) New non-angiogenesis dependent pathways for tumour growth. Eur J Cancer 39:1835–1841

Ribatti D, Finato N, Crivellato E et al (2005) Neovascularization and mast cells with tryptase activity increase simultaneously with pathological progression in human endometrial cancer. Am J Obstet Gynecol 193:1961–1965

Ribatti D, Ponzoni M (2005) Antiangiogenic strategies in neuroblastoma. Cancer Treat Rev 31:27–34

Ribatti D, Vacca A (2005) Therapeutic renaissance of thalidomide in the treatment of haematological malignancies. Leukemia 19:1523–1531

Ribatti D, Nico B, Vacca A (2006a) Importance of the bone marrow microenvironment in inducing the angiogenic response in multiple myeloma. Oncogene 25:4257–4266

Ribatti D, Vacca A, Nico B et al (2006b) Angiogenesis and anti-angiogenesis in hepatocellular carcinoma. Cancer Treat Rev 32:437–444

Ribatti D, Vacca A, Rusnati M et al (2007) The discovery of basic fibroblast growth factor/fibroblast growth factor-2 and its role in haematological malignancies. Cytokines Growth Fact Rev 18:327–334

Ribatti D (2009) Chick embryo chorioallantoic membrane as a useful tool to study angiogenesis. Int Rev Cell Mol Biol (in press)

Ridell, Norrby K (2001) Intratumoral microvascular density in malignant lymphomas of B-cell origin. APMIS 109:66–72

Romagnani P, Annunziato F, Lasagni L et al (2001) Cell cycle-dependent expression of CXC chemokines receptor 3 by endothelial cells mediates angiostatic activity. J Clin Invest 107:53–63

Romagnani P, Annunziato F, Liotta F et al (2005) CD14$^+$ CD34 low cells with stem cell phenotypic and functional features are the major source of circulating endothelial progenitors. Circ Res 97:314–322

Rondoni P (1946) Il cancro, Casa Editrice Ambrosiana, Milano

Ruoslahti E (2002) Specialization of tumour vasculature. Nat Rev Cancer 2002; 2:83–90

Sabin FR (1902) On the origin of the lymphatic system from the veins and the development of the lymph hearts and thoracic duct in the pig. Am J Anat 1:367–391

Saharinien P, Tammela T, Karkakainen MJ et al (2004) Lymphatic vasculature: development, molecular regulation and role in tumor metastasis and inflammation. Trends Immunol 25:387–395

Salgado R, Benoy I, Vermeulen P et al (2004) Circulating basic fibroblast growth factor is partly derived from the tumour in patients with colon, cervical and ovarian cancer. Angiogenesis 7:29–32

Salven P, Mustjoki S, Alitalo R et al (2003) VEGFR-3 and CD133 identify a population of CD34+ lymphatic/vascular endothelial precursor cells. Blood 101:168–72

Sandison JC (1928) Observations on growth of blood vessels as seen in transparent chamber introduced into rabbit's ear. Am J Anat 41:475–496

Sandlund J, Hedberg Y, Bergh A et al (2006) Endoglin (CD105) expression in human renal cell carcinoma. Br J Urol Int 97:706–710

Santamaria M, Moscatelli G, Viale GL et al (2003) Immunoscintigraphic detection of the ED-B domain of fibronectin, a marker of angiogenesis, in patients with cancer. Clin Cancer Res 9:571–579

Satchi-Fainaro R Mamluk R Wang L et al (2005) Inhibition of vessel permeability by TNP-470 and its polymer conjugate, caplostatin. Cancer Cell 7:251–261

Sato TN, Quin Y, Kozak CA et al (1993) Tie-1 and Tie-2 define another class of putative receptor tyrosine kinase genes expressed in early embryonic vascular system. Proc Natl Acad Sci USA 90:9355–9358

Sato TN, Tazawa Y, Deutsch U et al (1995) Distinct roles of the receptor tyrosine kinase Tie-1 and Tie-2 in blood vessel formation. Nature 376:70–74

Sawatsubashi M, Yamada T, Fukushima N et al (2000) Association of vascular endothelial growth factor and mast cells with angiogenesis in laryngeal squamous cell carcinoma. Virchows Arch 436:243–248

Scapini P, Nesi L, Morini M et al (2002) Generation of biologically active angiostatin kringle 1–3 by activate human neutrophils. J Immunol 168:5798–5804

Scavelli C, Nico B, Cirulli T et al (2008) Vasculogenic mimicry by bone marrow macrophages in patients with multiple myeloma. Oncogene 27:663–674

Schatteman GC, Hanlon HD, Jiao G et al (2000) Blood-derived angioblasts accelerate blood-flow restoration in diabetic mice. J Clin Invest 106:571–578

Schoppmann SF, Birner P, Stockl J et al (2002) Tumor-associated macrophages express lymphatic endothelial growth factors and are related to peritumoral lymphangiogenesis. Am J Pathol 161:947–956

Schwartz JD, Monea S, Marcus SG et al (1998) Soluble factor(s) released from neutrophils activates endothelial cell matrix metalloproteinase-2. J Surg Res 76:79–85

Senger DR, Galli SJ, Dvorak AM et al (1983) Tumor cells secrete a vascular permeability factor that promotes accumulation of ascites fluid. Science 219:983–985

Senger DR, van de Water L, Brown LF et al (1993) Vascular permeability factor (VPF, VEGF) in tumor biology. Cancer Metastasis Rev 12:303–324

Shaked Y, Emmenegger U, Man S et al (2005) Optimal biologic dose of metronomic chemotherapy regimens is associated with maximum antiangiogenic activity. Blood 106:3058–3061

Shaked Y, Ciarrocchi A, Franco M et al (2006) Therapy-induced acute recruitment of circulating endothelial progenitor cells to tumors. Science 313:1785–1787

Shamamian P, Schwartz JD, Pocock BJ et al (2001) Activation of progelatinase A (MMP-2) by neutrophil elastase, cathepsin G, and proteinase-3: a role for inflammatory cells in tumor invasion and angiogenesis. J Cell Physiol 189:197–206

Shanafelt TD, Kay NE (2006) The clinical and biologic importance of neovascularization and angiogenic signaling pathways in chronic lymphocytic leukemia. Semin Oncol 33:174–185

Sharma S, Sharma MC, Gupta DK et al (2006) Angiogenic patterns and their quantitation in high grade astrocytic tumors. J Neurooncol 79:19–30

Sharpe RJ, Byers HR, Scott CF et al (1990) Growth inhibition of murine melanoma and human colon carcinoma by recombinant human platelet factor 4. J Natl Cancer Inst 82:848–853

Shi Q, Rafii S, Wu MHD et al (1998) Isolation of putative progenitor endothelial cells for angiogenesis. Blood 92:362–367

Shimizu T, Hino K, Tauchi K et al (2000) Prediction of axillary lymph node metastasis by intravenous digital subtraction angiography in breast cancer, its correlation with microvascular density. Breast Cancer Res Treat 61:261–269

Shing Y, Folkman J, Sullivan R et al (1984) Heparin-affinity: purification of a tumor-derived capillary endothelial cell growth factor. Science 223:1296–1299

Shing Y, Folkman J, Haudenschild C et al (1985) Angiogenesis is stimulated by a tumor-derived endothelial cell growth factor. J Cell Biochem 29:275–287

Shirakawa K, Shibuya M, Heike Y et al (2002) Tumor-infiltrating endothelial cells and endothelial precursor cells in inflammatory breast cancer. Int J Cancer 99:344–351

Sica A, Schioppa T, Mantovani A et al (2006) Tumour-associated macrophages are a distinct M2 polarized population promoting tumour progression: potential targets of anti-cancer therapy. Eur J Cancer 42:717–727

Sidky YA, Borden EC (1987) Inhibition of angiogenesis by interferons: effects on tumor- and lymphocyte-induced vascular responses. Cancer Res 47:5155–5161

Singh RK, Gutman M, Bucana CD et al (1995) Interferons α and β down-regulate the expression of basic fibroblast growth factor in human carcinomas. Proc Natl Acad Sci USA 92:4562–4566

Singhal S, Mehta J, Desikan R et al (1999) Antitumor activity of thalidomide in refractory multiple myeloma. N Engl J Med 341:1565–1571

Sippola-Thiele M, Hanahan D et al (1988) Cell heritable stages of tumor progression in transgenic mice harboring the bovine papilloma virus type 1 genome. Mol Cell Biol 9:925–934

Skobe M, Hamberg LM, Hawighorst T et al (2001a) Concurrent induction of lymphangiogenesis, angiogenesis, and macrophage recruitment by vascular endothelial growth factor-C in melanoma. Am J Pathol 159:893–903

Skobe M, Hawighorts T, Jackson DG et al (2001b) Induction of tumor lymphangiogenesis by VEGF-C promotes breast cancer metastasis. Nat Med 7:192–198

Slaton JW, Perrott EP, Inoue K et al (1999) Interferon-α mediated down-regulation of angiogenesis-related genes and therapy of bladder cancer are dependent on optimization of biological dose and schedule. Clin Cancer Res, 5:2726–2734

Smith JK, Mamoon NM, Duher RJ (2004) Emerging roles of targeted small molecule protein-tyrosine kinase inhibitors in cancer therapy. Oncol Res 14:175–225

Smith-Mc Cune K, Zhu YH et al (1997) Cross-species comparisons of angiogenesis during the premalignant stages of squamous carcinogenesis in the human cervix and K14-HOV16 transgenic mice. Cancer Res 57:1294–1300

Soker S, Takashima S, Miao HQ et al (1998) Neuropilin-1 is expressed by endothelial and tumor cells as an isoform-specific receptor to vascular endothelial growth factor. Cell 92:735–745

Solovey A, Lin Y, Browne P et al (1997) Circulating activated endothelial cells in sickle cell anemia. N Engl J Med 337:1584–1590

Song S, Ewald AJ, Stallcup W et al (2005) PDGFRbeta+ perivascular progenitor cells in tumours regulate pericyte differentiation and vascular survival. Nat Cell Biol 7:870–879

Sood AK, Seftor EA, Fletcher MS et al (2001) Molecular determinants of ovarian cancer plasticity. Am J Pathol 158:1279–1288

Spitz DR (2000) Glucose deprivation-induced oxidative stress in human tumor cells. Ann NY Acad Sci 899:349–362

Spurbeck WW, Ng CY, Strom TS et al (2002) Enforced expression of tissue inhibitor of matrix metalloproteinase-3 affects functional capillary morphogenesis and inhibits tumor growth in a murine tumor model. Blood 100:3361–3368

Srivastava A, Laidler P, Davies R et al (1988) The prognostic significance of tumor vascularity in intermediate-thickness (0.76–4.0 mm thick) skin melanoma. Am J Pathol 133:419–423

St Croix B, Rago C, Velculescu V et al (2000) Gene expressed in human tumor and endothelium. Science 289:1197–1202

Stacker SA, Baldwin ME, Achen MG (2002) The role of tumor lymphangiogenesis in metastatic spread. FASEB J 16:922–934

Starkey JR, Crowle PK, Taubenberger S (1988) Mast-cell-deficient W/W^v mice exhibit a decreased rate of tumor angiogenesis. Int J Cancer 42:48–52

Strasly M, Cavallo F, Guena M et al (2001) IL-12 inhibition of endothelial cell functions and angiogenesis depends on lymphocyte-endothelial cell cross-talk. J Immunol 166:3890–3899

Straubel B, Chott A, Huber D et al (2004) Lymphoma-specific genetic aberrations in microvascular endothelial cells in B-cell lymphomas. N Engl J Med 351:250–259

Steinberg F, Rohrborn HJ, Otto T et al (1997) NIR reflection measurements of hemoglobin and cytochrome aa3 in healthy tissue and tumors. Correlation to oxygen consumption: preclinical and clinical data. Adv Exp Med Biol 428:69–77

Strander H (1986) Interferon treatment of human neoplasia. Adv Cancer Res 46:1–265.

Stratmann A, Risau W, Plate KH (1998) Cell-type specific expression of angiopoietin-1 and angiopoietin-2 suggests a role in glioblastoma angiogenesis. Am J Pathol 153:1459–1466

Strehlow K, Werner N, Berweiler J et al (2003) Estrogen increases bone marrow-derived endothelial progenitor cell production and diminishes neointima formation. Circulation 107:3059–3065

Sund M, Hanano Y, Sugimoto H et al (2005) Function of endogenous inhibitors of angiogenesis as endothelium-specific tumor suppressors. Proc Natl Acad Sci USA 102:2934–2939

Sunderkotter C, Goebeler M, Schiltze-Osthoff K et al (1991) Macrophage-derived angiogenesis factors. Pharmacol Ther 51:195–216

Suri C, Jones PF, Patan S et al (1996) Requisite role of angiopoietin-1, a ligand for the TIE2 receptor, during embryonic angiogenesis. Cell 87:1171–1180

Sweeney CJ, Miller KD, Sledge GW Jr (2003) Resistance in the anti-angiogenic era: nay-saying or a word of caution? Trends Mol Med 9:24–29

Takahashi T, Kalka C, Masuda H et al (1999) Ischemia and cytokine-induced mobilization of bone marrow-derived endothelial progenitor cell for neovascularization. Nat Med 5:434–438

Takanami I, Takeuchi K, Narume M (2000) Mast cell density is associated with angiogenesis and poor prognosis in pulmonary adenocarcinoma. Cancer 88:2686–2692

Talks KL, Turley H, Gatter KC et al (2000) The expression and distribution of the hypoxia-inducible factors HIF-1 α and HIF-1 β in normal human tissues, cancer, and tumor-associated macrophages. Am J Pathol 157:411–421

Tanaka S, Mori M, Sakamoto Y et al (1999) Biologic significance of angiopoietin-2 expression in human hepatocelluar carcinoma. J Clin Invest 103:341–345

Tannock IF (1968) The relation between cell proliferation and the vascular system in a transparent mouse mammary tumour. Br J Cancer 22:258–273

Taraboletti G, Roberts D, Liotta LA et al (1990) Platelet thrombospondin modulates endothelial cell adhesion, motility, and growth: a potential angiogenesis regulatory factor. J Cell Biol 111:765–772

Taylor S, Folkman J (1982) Protamine is an inhibitor of angiogenesis. Nature 297:307–312

Tee MK, Vigne JL, Taylor RN (2006) All-trans retinoic acid inhibits vascular endothelial growth factor expression in a cell model of neutrophil activation. Endocrinology 147:1264–1270

Teruya-Feldstein J, Jaffe ES, Burd PR et al (1999) Differential chemokine expression in tissues involved by Hodgkin's disease: direct correlation of eotaxin expression and tissue eosinophilia. Blood 93:2463–2470

Theoharides TC, Conti P (2004) Mast cells: the JEKYLL and HYDE of tumor growth. Trends Immunol 25:235–241

Thomlinson RH, Gray LH (1955) The histological structure of some human lung cancers and the possible implications for radiotherapy. Br J Cancer 9:539–549

Tomita M, Matsuzaki Y, Onitsuka T (2000) Effect of mast cell on tumor angiogenesis in lung cancer. Ann Thorac Surg 69:1686–1690

Toth T, Toth-Jakatics R, Jimi S et al (2000) Cutaneous malignant melanoma: correlation between neovascularization and peritumoral accumulation of mast cells overexpressing vascular endothelial growth factor. Human Pathol 31:955–960

Tozer GM, Kanthou C, Baguley BC (2005) Disrupting tumour blood vessels. Nat Rev Cancer 5:423–435

Traweek ST, Kandalaft PL, Mehta P et al (1991) The human hematopoietic progenitor cell antigen (CD34) in vascular neoplasia. Am J Clin Pathol 96:25–31

Trinchieri G (1993) Interleukin-12 and its role in the generation of Th1 cells. Immunol Today 14:335–338

Tuan D, Smith S, Folkman J et al (1973) Isolation of the nonhistone proteins of rat Walker 256 carcinoma. Their association with tumor angiogenesis. Biochemistry 12:3159–3165

Turner HE, Nagy Z, Gatter KC et al (2000) Angiogenesis in pituitary adenomas and the normal pituitary gland. J Clin Endocrinol Metab 85:1159–1162

Udagawa T, Fernandez A, Achilles EG et al (2002) Persistence of microscopic human cancers in mice: alterations in the angiogenic balance accompanies loss of tumor dormancy. FASEB J 16:1361–1370

Urbich C, Heeschen C, Aicher A et al (2003) Relevance of monocytic features for neovascularization capacity of circulating endothelial progenitor cells. Circulation 108:2511–2516

Vacca A, Ribatti D, Roncali L et al (1994) Bone marrow angiogenesis and progression in multiple myeloma. Br J Haematol 87:503–508

Vacca A, Moretti S, Ribatti D et al (1997) Progression of mycosis fungoides is associated with changes in angiogenesis and expression of the matrix metalloproteinase-2 and –9. Eur J Cancer 33:1685–1692

Vacca A, Ribatti D, Presta M et al (1999) Bone marrow neovascularization, plasma cell angiogenic potential, and matrix metalloproteinase-2 secretion parallel progression of human multiple myeloma. Blood 93:3064–3073

Vacca A, Ria R, Semeraro F et al (2003) Endothelial cells in the bone marrow of patients with multiple myeloma. Blood 102:3340–3348

Vacca A, Ribatti D (2006) Bone marrow angiogenesis in multiple myeloma. Leukemia 20:193–199

Vajkoczy P, Farhadi M, Gaumann A et al (2002) Microtumor growth initiates angiogenic sprouting with simultaneous expression of VEGF, VEGF receptor-2, and angiopoietin-2. J Clin Invest 109:777–785

Vajkoczy P, Blum S, Lamparter M et al (2003) Multistep nature of microvascular recruitment of ex vivo-expanded embryonic endothelial progenitor cells during tumor angiogenesis. J Exp Med 197:1755–1765

Van de Rijn M, Rouse RV (1994) CD34: a review. Appl Immunohistochem 2:71–80

Vasa M, Fitchtlscherer S, Aicher A et al (2001) Number and migratory activity of circulating progenitor cells inversely correlate with risk factors for coronary artery disease. Circ Res 89:E1–E7

Veikkola T, Jussila L, Makinen T et al (2001) Signalling via vascular endothelial growth factor receptor-3 is sufficient for lymphangiogenesis in transgenic mice. Embo J 20:1223–1231

Venneri MA, De Palma M, Ponzoni M et al (2007) Identification of proangiogenic Tie-2 expressing monocytes (TEMs) in human peripheral blood and cancer. Blood 109:5276–5285

Vermeulen PB, Verhoeven D, Hubens G et al (1995) Microvessels density, endothelial cell proliferation and tumor cell proliferation in human colorectal adenocarcinomas. Ann Oncol 6:59–64

Vermeulen PB, Gasparini G, Fox SB et al (1996) Quantification of angiogenesis in solid human tumors: an international consensus on the methodology and criteria of evaluation. Eur J Cancer 32A:2474–2484

Vlodavsky I, Korner G, Ishai-Michaeli R et al (1990) Extracellular matrix-resident growth factors and enzymes: possible involvement in tumor metastasis and angiogenesis. Cancer Metastasis Rev 9:203–226

Vogelstein B (1993) The multistep nature of cancer. Trends Genet 9:138–141

Volm M, Koomagi R, Mattern J et al (1997) Prognostic value of basic fibroblast growth factor and its receptor (FGFR-1) in patients with non-small cell lung carcinomas. Eur J Cancer 33:691–693.

Vu TH, Werb Z (2000) Matrix metalloproteinases: effectors of development and normal physiology. Genes Dev 14:2123–2133

Wang JM, Kumar S, Pye D et al (1993) A monoclonal antibody detects heterogeneity in vascular endothelium of tumors and normal tissues. Int J Cancer 54:363–370

Wang JM, Kumar S, Pye D et al (1994) Breast carcinoma: comparative study of tumor vasculature using two endothelial cell markers. J Natl Cancer Inst 86:386–388

Warren BA (1979) The vascular morphology of tumors. In: Patherson HI (ed) Tumor blood circulation: angiogenesis, vascular morphology and blood flow of experimental and human tumors, CRC Press Inc, Boca Raton, pp 1–48

Warren RS, Yuan H, Matli MR et al (1995) Regulation by vascular endothelial growth factor of human colon cancer tumorigenesis in a mouse model of experimental liver metastasis. J Clin Invest 95:1789–1797

Watanabe Y, Lee SW, Detmar M et al (1997) Vascular permeability factor/vascular endothelial growth factor (VEGF) delays and induces escape from senescence in human dermal microvascular endothelial cells. Oncogene 14:2025–2032

Watnick RS, Cheng YN, Rangarajan A et al (2003) Ras modulates Myc activity to repress thrombospondin-1 expression and increase tumor angiogenesis. Cancer Cell 3:219–231

Weidner N, Sample J, Welch W et al (1991) Tumor angiogenesis correlates with metastasis in invasive breast carcinoma. New Engl J Med 324:1–8

Weidner N, Carroll PR, Flax J et al (1993) Tumor angiogenesis correlates with metastasis in invasive prostate carcinoma. Am J Pathol 143:401–409

Weiss JB, Brown RA, Kumar S et al (1979) An angiogenic factor isolated from tumours: a potent low-molecular weight compound. Brit J Cancer 40:493–496

Wesseling P, Schlingemann RO, Reiveld FJ et al (1995) Early and extensive contribution of pericytes/vascular smooth muscle cells to microvascular proliferation in glioblastoma multiforme: an immuno-light and immuno-electron microscopic study. J Neuropathol Exp Neurol 54:304–310

Whalen GF (1990) Solid tumors and wounds: transformed cells misunderstood as injured tissues? Lancet 336:1489–1492

Whitaker GB, Limberg BJ, Rosembaum JS (2001) Vascular endothelial growth factor receptor-2 and neuropilin-1 form a receptor complex that is responsible for the differential signaling potency of VEGF(165) and VEGF(121) J Biol Chem 276:25520–25531

Whitelock JM, Murdoch AD, Iozzo RV et al (1996) The degradation of human endothelial cell-derived perlecan and release of bound basic fibroblast growth factor by stromelysin, collagenase, plasmin, and heparanases. J Biol Chem 271:10079–10086

Wigle JT, Oliver G (1999) Prox1 function is required for the development of murine lymphatic system. Cell 98:769–778

Williams AO, Ward JM, Jackson MA et al (1995) Immunohistochemical localization of transforming growth factor-beta1 in Kaposi's sarcoma. Human Pathol 26:469–473

Wilting J, Aref Y, Huang R et al (2006) Dual origin of avian lymphatics. Dev Biol 292:165–173

Wood JM, Bold G, Buchdunger E et al (2000) PTK 787/ZK222584, a novel and potent inhibitor of vascular endothelial growth factor receptor tyrosine kinases, impairs vascular endothelial growth factor responses and tumor growth after oral administration. Cancer Res 60:2178–2189

Wu Y, Zhong Z, Huber J et al (2006) Antivascular endothelial growth factor receptor-1 antagonist antibody as a therapeutic agent for cancer. Clin Cancer Res 12:6573–6584

Yancopoulos GD, Davis S, Gale NW et al (2000) Vascular specific growth factors and blood vessels formation. Nature 407:242–248

Yao L, Sgadari C, Furuke K et al (1999) Contribution of natural killer cells to inhibition of angiogenesis by interleukin-12. Blood 93:1612–1621

Yamagishi S, Yonekura H, Yamamoto Y et al (1999) Vascular endothelial growth factor acts as a pericyte mitogen under hypoxic conditions. Lab Invest 79:501–509

Yamaguchi J, Kusano KF, Masuo O et al (2003) Stromal cell-derived factor-1 effects on ex vivo expanded endothelial progenitor cell recruitment for ischemic neovascularization. Circulation 107: 1322–1328

Yang JC, Haworth L, Sherry RM et al (2003) A randomized trial of bevacizumab, an anti-VEGF antibody, for metastatic renal cancer. New Engl J Med 349:427–434

Yao X, Qian CN, Zhang ZF et al (2007) Two distinct types of blood vessels in clear cell renal cell carcinoma have contrasting prognostic implications. Clin Cancer Res 13:161–169

Yang Q, Rasmussen SA, Friedman JM (2002) Mortality associated with Down's syndrome in the USA from 1983 to 1997: a population-based study. Lancet 359:1019–1025

Yiangou C, Gomm JJ, Coope RC et al (1997) Fibroblast growth factor 2 in breast cancer: occurrence and prognostic significance. Br J Cancer 75:28–33

Yoshii Y, Sugiyama K (1988) Intercapillary distance in the proliferation area of human glioma. Cancer Res 48:2938–2941

Yin AH, Miraglia S, Zanjani ED et al (1997) AC133, a novel marker for human hematopoietic stem and progenitor cells. Blood 90:5002–5012

Yu JL, Rak JW, Carmeliet P et al (2001) Heterogeneous vascular dependence of tumor cell populations. Am J Pathol 158:1325–1334

Yu JL, Rak JW, Coomber BL et al (2002) Effect of p53 status on tumor response to antiangiogenic therapy. Science 295:1526–1528

Yuan F, Chen Y, Dellian M et al (1996) Time-dependent vascular regression and permeability changes in established xenografts induced by an anti-vascular endothelial growth factor/vascular permeability factor antibody. Proc Nat Acad Sci USA 93:14765–14770

Yuan L, Moyon D, Pardanaud L et al (2002) Abnormal lymphatic vessel development in neuropilin-2 mutant mice. Development 129:4797–4806

Yue WY, Chen ZP (2005) Does vasculogenic mimicry exist in astrocytoma? J Histochem Cytochem 53:997–1002

Zagzag D, Hooper A, Friedlander DR et al (1999) In situ expression of angiopoietins in astrocytomas identifies angiopoietin-2 as an early marker of tumor angiogenesis. Exp Neurol 159:391–400

Zhang S, Zhang D, Sun B (2007) Vasculogenic mimicry: Current status and future prospects. Cancer Lett 254:157–164.

Zhang X, Ibrahimi OA, Olsen SK et al (2006) Receptor specificity of the fibroblast growth factor family. The complete mammalian FGF family. J Biol Chem 281:15694–15700

Zhao ZS, Zhou JL, Yao GY et al (2005) Correlative studies on bFGF mRNA and MMP-9 mRNA expressions with microvascular density, progression, and prognosis of gastric carcinomas. World J Gastroenterol 11:3227–3233

Zhong H, Bowen JP (2006) Antiangiogenesis drug design: multiple pathways targeting tumor vasculature. Curr Med Chem 13:849–862

Zorick TS, Mustacchi Z, Bando SY et al (2001) High serum endostatin levels in Down syndrome: implications for improved treatment and prevention of solid tumors. Eur J Hum Genet 9:811–814

Printed in the United States
134747LV00002B/107/P